WILD WORDS

AN ILLUSTRATED DICTIONARY OF SCIENTIFIC TERMS

BY MEGHAN LINDSAY

Australian Geographic

CONTENTS

Krefft's sugar glider is ARBOREAL.

AUTHOR'S INTRODUCTION

The monarch butterfly (above) was INTRODUCED to Australia.

The cockatiel or quarrion (left) is ENDEMIC to Australia.

WORDS SHAPE OUR WORLD.
They're also a fundamental part of science. They help us communicate with each other and express ourselves. We use them to describe different animals and plants, their appearance, and their behaviour. They allow us to categorise various types of rocks, periods in time, and even different types of clouds.

I've been fascinated by words ever since I was young. I loved reading books, and I remember sneaking to turn the lamp back on to keep reading after I was meant to go to sleep. I was also fascinated by animals. I loved watching the lizards dart around the backyard, and I wrote school assignments about my favourite animals, like the orca.

As I got older, I remained enthralled by words, and not only by animals but by science in general. This is exactly the kind of book I would have loved to read when I was younger – a book packed with wonderful science words. It also includes plenty of awesome photos and fun facts! I got to include cool information – such as that the lyrebird moves more than 330 tonnes of dirt a year, and that the peacock jumping spider uses bright colours and wacky dance moves to attract females. Seriously, there are so many amazing science words that it wouldn't be possible to fit all of them in this book. But within these pages, you'll find some of the most incredible words, pictures and facts to take with you on your science journey.

– MEGHAN LINDSAY

MEGHAN LINDSAY *(right) is an environmental writer with a background in conservation, natural resource management and science communication.*

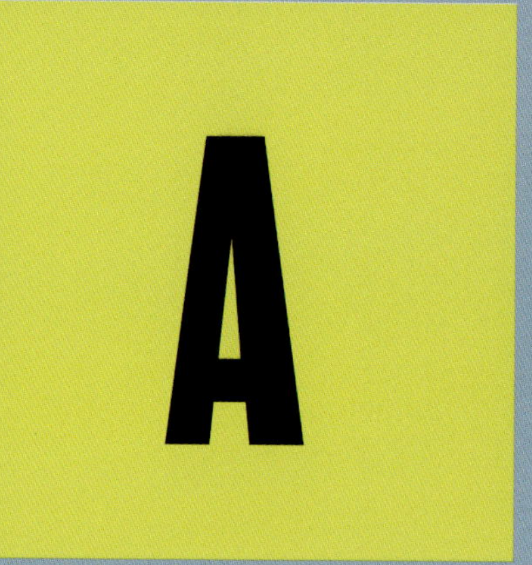

ABIOGENESIS
How living organisms have developed from non-living matter through chemical reactions, i.e. the origin of life. This is thought to have happened in stages, rather than as a single event.

ABIOTIC
Physical and chemical elements of an ecosystem that are not alive, such as rocks, sand, wind and sunlight. See also *biotic*.

ABORAL
The area on an animal that is away from, or on the opposite side to, the mouth. This term is used mostly for animals whose upper and lower sides are not distinct, such as echinoderms (e.g. sea stars). See also *adoral*.

ABSCISSION
When a plant sheds an organ that has grown old.

ABYSSAL ZONE
The deepest parts of the ocean, where the water is near freezing, there is no light at all, and the pressure is crushing; roughly 2000–6000 m deep. These areas account for about 75% of the ocean floor. See also *bathyal zone*.

Animals at this depth have special adaptations for living in this type of environment, such as **Sloane's viperfish** (*Chauliodus sloani*, left) and the pelican eel (*Eurypharynx pelecanoides*), which use bioluminescence to attract their prey.

Sloane's viperfish inhabits the ABYSSAL ZONE.

ACCLIMATISATION SOCIETY
These societies were created in the mid-1800s in Australia to introduce species from other countries. Species were either introduced for financial gain, to hunt, or simply to

The European carp (far left) and the red fox (left) were introduced by ACCLIMATISATION SOCIETIES.

make Australia feel more like Europe for homesick settlers. Unfortunately, this has led to serious pest problems and environmental damage, such as that caused by the **European carp** (*Cyprinus carpio,* top) in the Murray River, or by the **red fox** (*Vulpes vulpes*, top right), which preys on native mammals.

ACEPHALOUS
Animals that do not have a distinct head. The head of an oyster or a mussel is indistinguishable from its body, as these animals are acephalous.

ACICULAR
A term used in geology and botany to describe a sharp or needle-shaped structure. A long, sharp crystal and the needles of a pine tree are acicular.

ACULEATE
Organisms that are prickly, pointed or that have stingers.

The spider-hunting wasp, or **orange spider wasp** (*Cryptocheilus bicolor,* right) is aculeate and has one of the most painful stings of any insect in Australia. These wasps attack huntsman spiders and drag them back to their nests to feed to their babies!

ADAPTATION
When organisms change in response to their environment, either genetically or by changing their physiology or behaviour. This gives the organism a better chance of survival. See also *plasticity*.

The spider-hunting wasp (above) is ACULEATE.

About 18 species of Darwin's finch (above) in the Galapagos Islands evolved by ADAPTIVE RADIATION.

ADAPTIVE RADIATION

When many species evolve from a common ancestor and spread to new environments by adapting to different habitats.

ADELPHOPARASITE

A parasite that uses a host that is closely related. *Tikvahiella candida*, for example, is a type of marine-living adelphoparasite that attaches itself to the robust red seaweed (*Solieria robusta*).

ADIPOSE EYELID

A thick, transparent skin covering the eyes of some species of fish. These eyelids protect the eye from ultraviolet light and from being damaged by foreign objects.

ADORAL

To be on the same side of an organism as the mouth. See also *aboral*.

AEDEAGUS

The male sex organ of most groups of insects.

AEROBIC

An environment that has oxygen, an organism that requires oxygen, or a process that can only happen when there is oxygen available.

AEROTAXIS

When an organism changes the direction in which it is moving in response to the levels of oxygen in the environment.

AESTIVATION

Sometimes known as 'summer sleep', aestivation is when organisms become dormant or inactive in hotter weather, similar to hibernation but in warmer months. See also *hibernation, torpor*.

The western swamp tortoise (below) undergoes AESTIVATION.

The **western swamp tortoise** (*Pseudemydura umbrina,* left) is a critically endangered species that lives in Western Australia. The species was thought to be extinct for more than 100 years before it was rediscovered in 1953. In summer, when the weather is too warm, they aestivate under leaf litter, logs, or holes dug by other animals.

AGGRESSION

Any behaviour used by an animal to intimidate or injure another animal, not including behaviour related to predation.

Some animals have unique ways of showing aggression, such as the **ghost crab** (*Ocypode cordimanus*, above) found along most Australian shorelines. Not only does it use its powerful claws to show aggression, but it also growls at predators by grinding together tiny teeth in its stomach!

AGONISTIC BEHAVIOUR

Animal behaviour that relates to fighting but may include aggression, defence, submission and retreat.

ALARM RESPONSE

A signal from an animal to warn others of danger. These can include visual signals, calls, thumps and even smells. Australian pademelons, such as the **red-necked pademelon** (*Thylogale thetis*, pictured), 'thump' their hindlegs on the ground as an alarm response.

ALBINISM

A mutation that affects the production of melanin, a pigment responsible for the colour in eyes, skin, fur and leaves. Without melanin, animals and plants cannot produce their usual colours and will instead be pure white. See also *leucism*.

Migaloo, the humpback whale (*Megaptera novaeangliae)*, is a famous example of an albino animal. He has been sighted periodically off the coast of Australia since 1991. On Bruny Island, Tasmania, there is also a community of albino **Bennett's wallabies** (*Notamacropus rufogriseus,* left), in which the genetic mutation has become common.

ALLELE

One of two or more different types of genes that occur in the same location on a pair of chromosomes.

The **Gouldian finch** (*Chloebia gouldiae*, pictured) is an example of when different alleles code for the same trait. Depending on which combination of alleles an individual bird inherits from its parents, it will end up with a black-, red- or gold-coloured head.

The river red gum demonstrates ALLELOPATHY.

Myotis bats use a form of ALLOTHETIC navigation.

Koalas (above right) and wombats (right) are ALTRICIAL.

ALLELOPATHY

When a plant produces chemicals that affect the germination, growth, survival and reproduction of other plants nearby.

The **river red gum** (*Eucalyptus camaldulensis,* above) is the most widespread eucalypt in Australia. It produces allelopathic chemicals that prevent its own seedlings from growing nearby. This eucalypt requires a flood to wash the chemicals from the soil so that its seeds can germinate.

ALLOPATRIC SPECIATION

When populations of a species become geographically isolated by mountains, rivers or other physical barriers and evolve genetic differences over time as a result, eventually becoming a new species.

Australia's 17 or more rock-wallaby species represent allopatric speciation that occurred over time as their rocky homes were separated by roads, cleared pastoral land, or housing developments. See also *sympatric speciation*.

ALLOTHETIC

When an animal uses spatial cues from sight, sound or smell to orient and navigate. See also *idiothetic*.

ALTITUDE

A measurement of distance, usually in a vertical direction.

ALTRICIAL

When an animal is born helpless and requires special parental care. Australian marsupials are highly altricial. Once they are born, they require months of further development inside their mother's pouch or with carers if orphaned. Many baby birds are also altricial. See also *precocial*.

The humpback whale (above) exhibits ALTRUISM.

Insects are often found preserved in AMBER (below).

ALTRUISM

When the behaviour of one individual provides a benefit to another individual at a cost to itself.

The **humpback whale** (*Megaptera novaeangliae*, above) demonstrates altruistic behaviour. Individuals have been observed fending off orca (*Orcinus orca*) to stop them from attacking seals, other whales and even fish.

ALVEOLI

Small sacs with an intricate vascular network that aid gas exchange. The lungs of mammals are filled with these small sacs, helping them extract oxygen from the air they breathe.

AMBER

Fossilised tree sap. Sometimes, amber contains the preserved bodies of insects and plant matter that were caught in sticky sap before it was fossilised. Scientists can use these preserved specimens to learn more about the lives of prehistoric species.

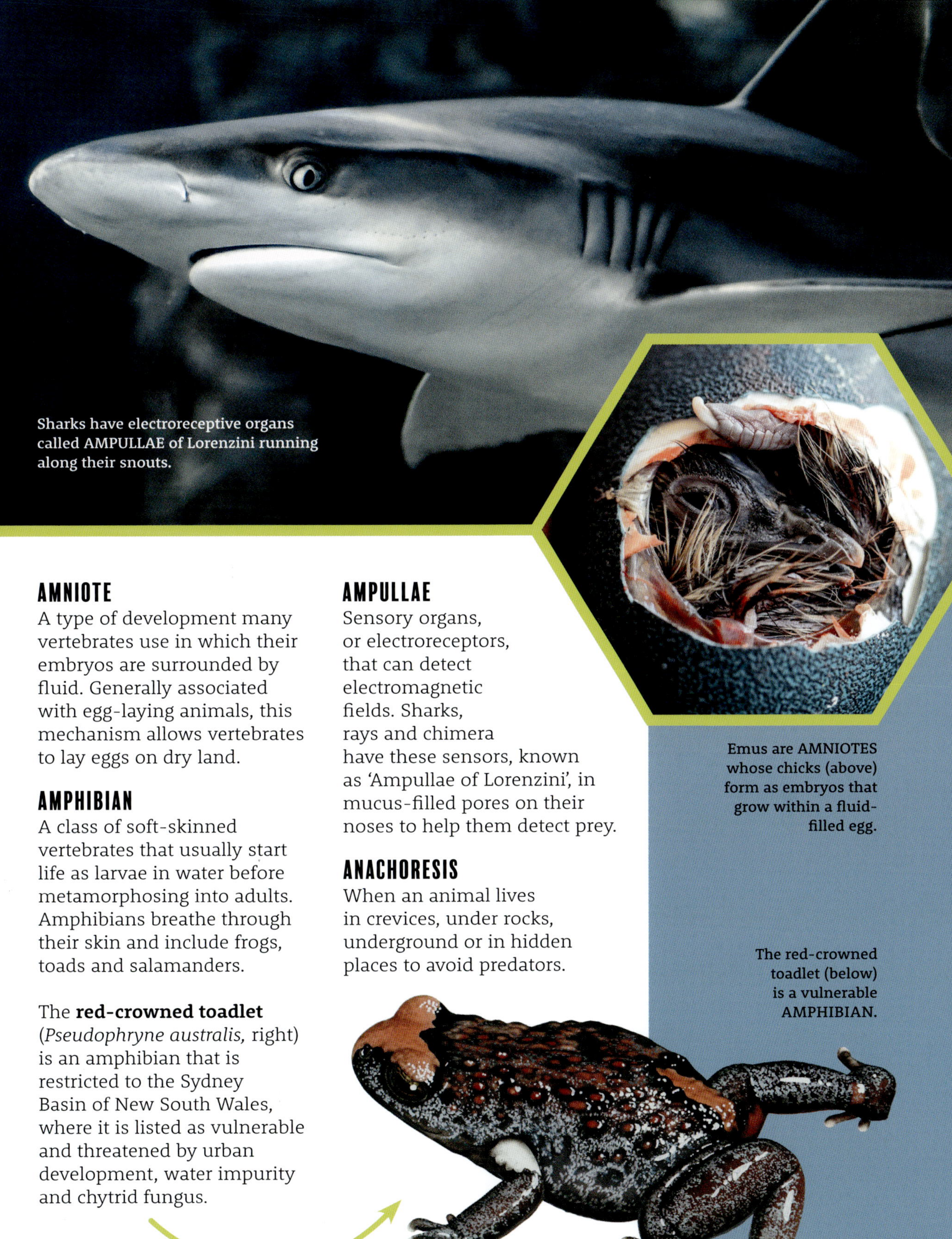

Sharks have electroreceptive organs called AMPULLAE of Lorenzini running along their snouts.

Emus are AMNIOTES whose chicks (above) form as embryos that grow within a fluid-filled egg.

The red-crowned toadlet (below) is a vulnerable AMPHIBIAN.

AMNIOTE

A type of development many vertebrates use in which their embryos are surrounded by fluid. Generally associated with egg-laying animals, this mechanism allows vertebrates to lay eggs on dry land.

AMPHIBIAN

A class of soft-skinned vertebrates that usually start life as larvae in water before metamorphosing into adults. Amphibians breathe through their skin and include frogs, toads and salamanders.

The **red-crowned toadlet** (*Pseudophryne australis,* right) is an amphibian that is restricted to the Sydney Basin of New South Wales, where it is listed as vulnerable and threatened by urban development, water impurity and chytrid fungus.

AMPULLAE

Sensory organs, or electroreceptors, that can detect electromagnetic fields. Sharks, rays and chimera have these sensors, known as 'Ampullae of Lorenzini', in mucus-filled pores on their noses to help them detect prey.

ANACHORESIS

When an animal lives in crevices, under rocks, underground or in hidden places to avoid predators.

Some types of rainbow trout are ANADROMOUS.

ANADROMOUS

A type of migration where fish that usually live in the sea, such as salmon, migrate into fresh water to reproduce.

The pouched lamprey (*Geotria australis*) can be found in south-western and south-eastern Australia. It lives in the ocean as a parasite feeding on other fish, but it migrates into freshwater rivers to reproduce. The **rainbow trout** (*Oncorhynchus mykiss,* above) also has some anadromous subspecies. See also *catadromous.*

ANAEROBIC

An environment that has no oxygen, an organism that can only exist in environments that do not have oxygen, or a process that can only happen when oxygen is not present.

ANATOMY

The structure of organisms, including their different parts.

ANGIOSPERM

Plants that use flowering as a form of reproduction and rely on animals and the wind for pollination, e.g. roses, tomatoes, banksias and eucalypts. See also *gymnosperm.*

ANIMAL

A eukaryotic, multicellular organism from the kingdom Animalia. Animals are generally heterotrophic, breathe air, are mobile, and reproduce sexually.

ANISODACTYL

Having four toes: three facing forwards and one backwards. See also *Passeriformes.*

ANTENNAE

A structure on insects, arthropods and crustaceans used for sensing. Different groups use it for various senses, including touch, air movement, heat, vibration, smell and taste. Some species have modified antennae that are used for mating, brooding or swimming. The antennae of snails, such as the Mitchell's rainforest snail (*Thersites mitchellae*), detect scents and orientation and sometimes even have eyespots. See also *ocellus.*

Snails (below) use their ANTENNAE for various purposes.

Banksias (right) and other flowering plants are ANGIOSPERMS.

ANTHOCYANINS give fruit and flowers colour (above).

ANTHOCYANINS

Soluble pigments that give fruit, flowers or leaves their red, purple or blue colours, which attract pollinators and protect against UV radiation.

ANTHROPOCENE

Another term for the late Holocene, the geological era we are currently in. The Anthropocene era is said to be defined by the substantial impact humans have had on the planet.

ANTHROPOMORPHISM

The act of attributing human characteristics to non-human objects or organisms.

ANTIGEN

A substance that may trigger an immune response when it enters an animal – for example, a particular molecule or a grain of pollen.

ANUS

In most animals, the exterior opening known as the anus is used to get rid of the remains of food that has not been absorbed. Some aquatic animals, such as the **Mary River turtle** (*Elusor macrurus,* right), which is also known as the bum-breathing turtle, use this opening to breathe!

The Mary River turtle (below) breathes from its ANUS.

The hibiscus harlequin bug is APOSEMATIC.

this system fails, it can lead to uncontrolled cell growth, which in turn can lead to the formation of tumours.

APHOTIC

An environment where there is no light.

APOPTOSIS

A type of programmed cell death in multicellular eukaryotes where damaged cells die to protect the organism. This is a necessary part of growth and development. When this system fails, it can lead to uncontrolled cell growth, which in turn can lead to the formation of tumours.

APOSEMATIC

Warning colours that signify an organism is venomous, poisonous or otherwise harmful are aposematic. Animals use these colours to avoid being attacked or eaten, warning predators they are dangerous. See also *Batesian mimicry, Müllerian mimicry*.

The **hibiscus harlequin bug** (*Tectocoris diophthalmus,* left), which can be found along the east coast and in northern Australia, is aposematic. Its shiny, iridescent colouration warns birds of this bug's chemical defences.

APOSTATIC SELECTION

A type of selection in which organisms have different forms within the same species, e.g. different colours, which can give them an evolutionary advantage.

The **common eastern froglet** (*Crinia signifera,* left) can be found across eastern Australia. Populations of these frogs have colour variations, with some colour morphs being less likely to be preyed upon in some habitats.

APPEASEMENT

Any behaviour by an animal that aims to reduce aggression by another member of the same species but does not include avoidance or escape.

APOSTATIC SELECTION has occurred in the common eastern froglet.

APPENDAGE

A part of an animal or plant that sticks out from the main body. Appendages are typically smaller and less important to the function of the individual. An arm, leg or branch, for example, is more expendable than a head or trunk.

AQUATIC

Organisms that live in water.

AQUIFER

Underground geological formations that hold water.

ARBOREAL

Animals that spend most of their time living in trees.

Australia has many arboreal mammal species, including the black-footed tree-rat (*Mesembriomys gouldii*), **greater glider** (*Petauroides* spp., right), eastern pygmy-possum (*Cercartetus nanus*) and the koala (*Phascolarctos cinereus*), which spends most of its life in the tree tops.

The greater glider is ARBOREAL and rarely comes down to the ground.

ARCHAEA

A group of micro-organisms that are similar to, but distinct from, bacteria. Many species live in extreme environments, such as in salt lakes, hot springs or geothermal vents at very high temperatures or pressures.

ARCHOSAURIAN

Members of a group of reptiles that includes dinosaurs, pterosaurs, crocodiles and birds.

ARTHROPOD

Joint-limbed invertebrates in the phylum Arthropoda, which includes insects, spiders, crabs, ticks and mites. An estimated 80% of the world's species are arthropods, and they can be found in every habitat on Earth. Many, such as mites, are tiny parasites that live in or on larger organisms. Arthropods have segmented bodies and typically display bilateral symmetry. Although most lay eggs, some (like scorpions) give birth to live young.

The native tarantula (left) is an ARTHROPOD.

An ASTEROID impact is thought to have wiped out the dinosaurs.

ASEXUAL REPRODUCTION

This occurs when an organism creates offspring without mating with another individual. There are different ways this can happen – for example, archaea, bacteria, and many anemones can split into two genetically identical individuals (a process also known as binary fission), some plants also grow new individuals from broken-off pieces of the parent plant (propagation), and some vertebrate reptiles or shark species can even reproduce without the need of a male. See also *parthenogenesis*, *sexual reproduction*.

ASTEROID

Rocky debris in space left over from the formation of the solar system. Asteroids have no atmosphere and vary greatly in size from about 10 m to 530 km across. See also *meteor*.

ARTIFICIAL SELECTION

Human-driven genetic selection where individual organisms are chosen for breeding based on having favourable traits. See also *natural selection* and *sexual selection*.

The red junglefowl (below) has undergone ARTIFICIAL SELECTION.

Humans have selectively bred chickens for thousands of years to produce the greatest number of eggs or the best meat. They look very different to the **red junglefowl** (*Gallus gallus*, left) from South-East Asia, from which they were originally bred.

ASTRONOMY

The study of objects that exist anywhere in the Universe beyond Earth's atmosphere.

ATMOSPHERE

One or more layers of gases surrounding a planet and held there by gravity. Earth's atmosphere is mostly made up of nitrogen and oxygen, with small amounts of other gases thrown in. Our atmosphere protects us from harmful solar radiation and warms the surface of the planet, allowing the formation of liquid water.

AURORA

A display of brightly coloured lights in the sky caused by particles released by the Sun interacting with Earth's upper atmosphere. Also known as polar lights, in the Southern Hemisphere, these lights are called the Aurora Australis or the Southern Lights. They are most visible in Antarctica, but they can sometimes be seen further north, including in parts of Tasmania and Victoria.

AUTOCHTHONOUS

Indigenous to, or formed within, the place where it was found or the system it is part of.

AUTOECIOUS

Parasitic organisms that complete their entire life cycle in one host species. See also *heteroecious*.

Cordyceps (*Ophiocordyceps unilateralis*) is a type of autoecious 'zombie ant fungus' that parasites on ants, spiders or other insects. *Cordyceps tenuipes* (right) is an autoecious parasite of moth or butterfly pupae and can be found in Australia.

Cordyceps fungi are AUTOECIOUS.

AUTONOMIC

Behavioural responses produced involuntarily by the nervous system, including sweating, panting, heart rate and digestion.

AUTOTOMY

Also known as self-amputation, autotomy is when an animal purposefully severs part of its own body, usually to avoid a predator. The part that is severed is then regrown. Many lizard species in Australia can do this, including larger lizards such as the **eastern blue-tongue** (*Tiliqua scincoides,* top right).

The wall lizard (above) and blue-tongue lizard (top right) use AUTOTOMY to drop a section of tail if threatened.

AUTOTROPH

Organisms that use carbon dioxide (CO_2) as their main or only source of energy. See also *chemosynthesis, heterotroph, photosynthesis*.

Eucalypts (left), are AUTOTROPHS that use CO_2 to create energy.

AVIFAUNA

Another word for birdlife or the sum of bird species in a place or ecosystem.

The white-faced heron (left) is part of Australia's AVIFAUNA.

AVOIDANCE

Behaviour that reduces an animal's exposure to danger, which can be either learned or innate.

B

BACTERIA

Organisms that are made up of a single cell. Bacteria were the first life forms on Earth. The thrombolites and stromatolites found at Lake Thetis, **Lake Clifton** (pictured) and Hamelin Pool in Western Australia are remnant, layered microbial reefs created by cyanobacteria known as blue-green algae millions of years ago.

The harlequin snake eel (below) uses BATESIAN MIMICRY to look like the banded sea krait (below right).

BASAL

In biological cladistics, basal refers to the oldest known ancestor in a family tree and thus the most primitive known species. For instance, basal extinct species of monotreme found in Australia suggest that this order evolved here.

BATESIAN MIMICRY

When one species of plant or animal mimics the bright colouring of a different animal that is unpalatable or poisonous in an attempt to protect themselves from predators. See also *Müllerian mimicry*.

The harmless **harlequin snake eel** (*Myrichthys colubrinus,* top) can be found in Australian waters. It mimics the black-and-white stripes of the venomous **yellow-lipped sea krait** (*Laticauda colubrina,* top right).

Anglerfish (left) are BATHYPELAGIC.

BATHYAL ZONE

A zone in the ocean above the abyssal zone. This dimly lit section at depths of 200–2000 m, is known as the 'twilight zone'. Many weird, wonderful and bioluminescent sea creatures inhabit this zone. See also *abyssal zone.*

BATHYPELAGIC

Marine life that lives in the dark, deep ocean at 1000–4000 m depth.

The crocodilefish (above) is a BENTHIC feeder that ambushes prey on the sea floor.

BEHAVIOURAL ECOLOGY

The study of an organism's behaviour in its natural environment. Behavioural ecology enables researchers and conservationists to help predict changes to environments when variable factors are altered.

BENTHIC

Living on the bottom of a body of water or associated with the bottom of a watercourse.

BILATERAL SYMMETRY

If you were to draw a line down a bilaterally symmetrical animal, both sides of the animal would mirror each other. Humans, as an example, are bilaterally symmetrical and so are orchid flowers and cycads. See also *radial symmetry*.

Cycads (above) and orchid flowers (left) are BILATERALLY SYMMETRICAL.

The Blue Mountains of New South Wales serve as a biogeographical barrier.

The greater sooty owl (right) has BINOCULAR VISION.

BINOCULAR VISION

When an animal can look at an object with both eyes at the same time, allowing it to judge the distance to an object. See also *monocular vision*. The southern boobook owl (*Ninox boobook*) and **greater sooty owl** (*Tyto tenebricosa*, left) have excellent binocular vision, which helps with depth perception while hunting at night.

BINOMIAL NOMENCLATURE

A system of scientifically naming and labelling species. In this system, two words are used, first the genus, and then the species epithet. The scientific name for humans is *Homo sapiens*. Several extinct species share the genus *Homo*, including *Homo heidelbergensis* of Africa and *Homo floresiensis* of Asia, as well as the Neanderthals (*Homo neanderthalensis*), which went extinct ~40,000 years ago.

BIOCENOLOGY

The study of natural communities and how members of a community interact with each other.

BIODEGRADABLE

Materials that can be easily broken down by living organisms and decomposers like bacteria and fungi.

BIODIVERSITY

The variety of species and their ecosystems.

BIOENERGETICS

In biology, the study of how energy flows and transforms between living organisms and the environment they inhabit.

BIOGEOGRAPHICAL BARRIER

A geographical barrier that prevents the movement and intermingling of species. Barriers can vary greatly, from mountain ranges and oceans, to variations in temperature or climate.

BIOGEOGRAPHY

The study of the geographic distribution of organisms and the factors that influence those distributions.

BIOLOGY

The study of living organisms.

BIOLUMINESCENCE

Living organisms that produce light from a chemical reaction.

Some species of bacteria – as well as many deep-sea fish, and even some fungi – are bioluminescent. The ghost fungus (*Omphalotus nidiformis*, right) is a toxic Australian mushroom that glows in the dark.

BIOLOGICAL CONTROL

Using organisms or pathogens to control invasive species.

The **cane toad** (*Rhinella marina*, left) is a toxic amphibian that was introduced to Australia in 1935 to eat beetle larvae, which were destroying sugar cane crops. Unfortunately, the cane toad ate a lot more than beetle larvae and soon became one of Australia's most invasive pests, responsible for poisoning many native species. Other biological control agents – such as the *Cactoblastis cactorum* moth which controls prickly pear – have been more successful, with strict research and control guidelines to prevent a second 'cane-toad' incident.

A common redshank forages following the BIORHYTHMS of the tides.

BIOMASS
The total mass of a group of organisms. For example, humans make up 0.01% of the total biomass of the planet, while plants make up 82.4%.

BIOME
A large area characterised by different features that include the soil, climate, plants and wildlife existing there. For example, aquatic, grassland, forest and desert.

BIOMETRICS
Applying statistical analysis to biological data to study unique individuals within a population.

BIONTIC
An adjective used to refer to an individual, as opposed to phyletic, which refers to a group. See also *phyletic*.

BIOPHILIA
The tendency of humans to be drawn to and want to connect with nature and other life forms.

BIORHYTHM
A cyclic or rhythmic biological process or function that is innate and present at birth. An example is wading shorebirds following the rhythms of the tides when foraging, or sleep and wake cycles in animals.

BIOSPHERE
All parts of the Earth where life exists.

BIOTA
All of the plants and animals that live in a particular area.

BIOTIC
Used to describe the living components of a system.

BIOTOXIN
A toxin that is biological in origin and is produced by a living thing.

BIOTURBATION

The movement of soil by plants and animals, for example by burrowing and digging. Bioturbation helps with many processes, including nutrient cycling, water filtration and plant germination.

While digging and searching for ants, the short-beaked echidna (*Tachyglossus aculeatus*) moves seven tonnes of soil every year – that's about eight trailer loads! This doesn't come close to the **superb lyrebird** (*Menura novaehollandiae*, left), which moves a massive 330 tonnes per year, or about 11 dump trucks worth. Superb lyrebirds hold the Guinness World Record for 'most surface material displaced by a land animal'.

Swamp wallabies (left) and giraffes (bottom left) are BROWSERS.

BLOOD (below) transports oxygen around an animal's body.

BLOOD
A fluid circulated through the body of animals that transports dissolved substances such as oxygen around the body.

BONE
The rigid tissue that makes up the skeleton of vertebrates, composed of different organic and inorganic compounds.

BOTANY
The study of plant life.

BRACKISH
Water with a level of salinity in between that of fresh water and sea water.

BRADYSPORY
When seeds are gradually released from a cone or a fruit.

BROWSER
A herbivorous animal that feeds on the leaves, shoots and fruits of high-growing plants, e.g. giraffes and some wallaby species. See also *grazer*.

BRUMATION
A type of inactive or sluggish state that ectotherms, such as snakes and lizards, enter during winter or cold weather. See also *aestivation*, *hibernation*, *torpor*.

BOTANY is the study of plant life.

Some skinks enter BRUMATION during winter.

BUCCAL INCUBATION

Also known as mouth brooding, this is when an animal incubates its eggs and guards its young in the mouth. A number of fish species do this, including the **yellowhead jawfish** (*Opistognathus aurifrons,* above), whose males incubate the eggs intraorally for 7–9 days. Australia was once also home to two species of gastric-brooding frog in the genus *Rheobatrachus*, which raised their broods in the stomach! Sadly, both species are now extinct.

BUTEONINE

Of or relating to hawks. Australia, of course, has no members of the genus *Buteo*, known as the short-winged hawks. Australia's hawks belong to the Accipitridae family and are classified into different genera. The black-breasted buzzard (*Hamirostra melanosternon*) of Australia occupies a different genus to the **common buzzard** (*Buteo buteo*, pictured) of Europe and Asia and is more closely related to the square-tailed kite (*Lophoictinia isura*).

CAMBRIAN

An era that describes the time period from about 542–488 million years ago. The Cambrian was an important period during which multicellular organisms, such as sea sponges, became more common and the first species of plants adapted to living on land.

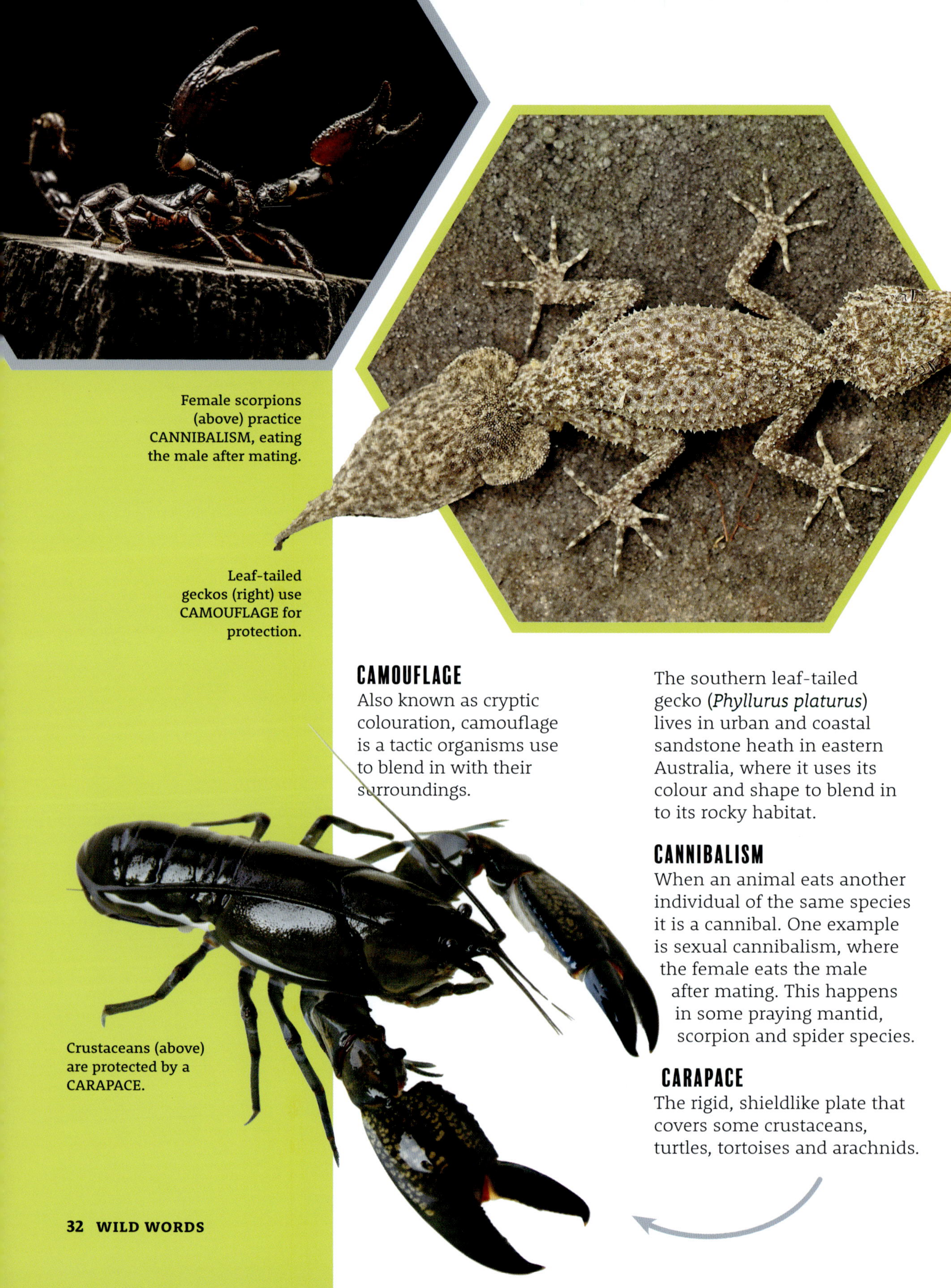

Female scorpions (above) practice CANNIBALISM, eating the male after mating.

Leaf-tailed geckos (right) use CAMOUFLAGE for protection.

Crustaceans (above) are protected by a CARAPACE.

CAMOUFLAGE

Also known as cryptic colouration, camouflage is a tactic organisms use to blend in with their surroundings.

The southern leaf-tailed gecko (*Phyllurus platurus*) lives in urban and coastal sandstone heath in eastern Australia, where it uses its colour and shape to blend in to its rocky habitat.

CANNIBALISM

When an animal eats another individual of the same species it is a cannibal. One example is sexual cannibalism, where the female eats the male after mating. This happens in some praying mantid, scorpion and spider species.

CARAPACE

The rigid, shieldlike plate that covers some crustaceans, turtles, tortoises and arachnids.

Australian sea lions (left) are CARNIVORE pinnipeds in the order Carnivora.

The southern shortfin eel is CATADROMOUS.

CARNIVORE

Animals that eat other animals as their primary food source, e.g. meat-eaters like lions and cats. This is also a generic term for animals in the order Carnivora. See also *folivore, frugivore, herbivore, omnivore*.

Apart from seals and sea-lions, Australia had no true carnivores (placental mammals in the order Carnivora) until the arrival of the dingo (*Canis lupis dingo*). Instead, this continent's meat-eaters were carnivorous marsupials. The largest known carnivorous mammal in Australia was the marsupial lion (*Thylacoleo carnifex*), which became extinct more than 30,000 years ago. When Europeans arrived in Australia, the largest meat-eating marsupial was the **Tasmanian tiger** (*Thylacinus cynocephalus*, below), which was confined to Tasmania and, sadly, went extinct in 1936.

CASQUE

In zoology, a casque is a hard, bony, helmet-like anatomical structure on the heads of some reptiles and birds, including the southern cassowary. The casque's purpose is yet to be determined.

CATADROMOUS

Migration in which freshwater fish migrate to the ocean in order to reproduce. A number of Australian fish species do this, including the Australian bass (*Percalates novemaculeata*) and the **southern shortfin eel** (*Anguilla australis*).

The chambered nautilus (left) is a CEPHALOPOD.

CELL
The individual units that make up life forms.

CEPHALOPODS
Marine molluscs with bilateral symmetry and prehensile tentacles.

CHEMICAL
A substance that can exist as a solid, liquid or gas and may change between these states at different temperatures.

CHEMOSYNTHESIS
The process of using inorganic compounds as a source of energy. Some bacteria use this process to produce food. See also *photosynthesis*.

CHEMOTAXIS
When an organism changes the direction in which it is moving in response to the levels of a particular chemical in the environment.

CHITIN
The rigid substance that forms the cuticles of arthropods.

CHLOROPHYLL
A green pigment used for photosynthesis that can be found in plants.

CHROMATOPHORE
A cell in the skin of some animals that contains colour pigments, allowing the animal to change colour. The **giant cuttlefish** (*Sepia apama*, right) is an Australian cephalopod. Chromatophores enable it to change colour in an instant for camouflage or mating displays. These cuttlefish undertake an annual mass migration to the Spencer Gulf, SA, to reproduce.

CHRYSALIS
The fragile outer coating a butterfly pupa lives in temporarily during its life cycle.

A CHRYSALIS (right) is the hard outer case of a butterfly pupa.

Giant cuttlefish have CHROMATOPHORES that help them change colour.

Wispy CIRRUS clouds over Lake Macquarie, NSW.

CIRCADIAN RHYTHM

The changes in patterns of activity that happen in animals over a 24-hour period. These changes are often, but not always, linked to changes in light and dark.

CIRRUS CLOUDS

Short, detached wispy clouds made from ice crystals and found at high altitudes.

CLADE

A grouping of organisms that is based on their hypothetical evolutionary history, rather than their physical features or appearance. Clades usually branch off from a single phylum and create an evolutionary tree showing a presumed common ancestor and its descendents.

Not all organisms fit neatly into a monophyletic CLADISTIC system. For example, scientists are still figuring out the relationships between sea jellies, ctenophores and salps.

CLADISTICS

A type of classification based on evolutionary traits and ancestry. Cladistics separates phyla into 'clades' based on their most likely ancestors and creates cladograms (or evolutionary trees) to show the relationships between them.

CLIMATE

The average patterns of weather experienced in an area over a long period of time.

CLIMATE CHANGE

Long-term changes in the average patterns of weather. Natural cycles can lead to changes in climate. More recently, human activity has also altered the climate. If species can't adapt to rapid changes, they risk becoming extinct. See also *atmosphere*.

The first mammal to go extinct due to climate change was the Bramble Cay melomys (*Melomys rubicola*). This small rodent lived on a tiny island in the Torres Strait. By 2014, due to the impacts of rising sea levels and dramatic weather events following climate change, there were no individuals left.

CLOACA

In many vertebrates, the cloaca is a single exterior opening used for reproduction and excreting food waste and urine.

CNIDARIANS

Aquatic, radially symmetrical animals with tentacles that include corals, sea jellies (left) and sea anemones.

COMMENSALISTIC

A type of symbiotic relationship between two organisms that benefits one of the species but does not disadvantage the other. See also *mutualistic, mycorrhizae, parasitic, symbiotic*.

Cattle egrets have a COMMENSALISTIC relationship with cows and buffalo.

The **cattle egret** (*Bubulcus ibis*, top right) follows herds of buffalo to eat the insects stirred up by their grazing. Bovines don't benefit, but if the birds pick off and eat ticks, the relationship becomes more *mutual* than *commensal*.

COMPETITION

When individuals of the same species or different species interact and both are negatively affected. This happens when individuals depend on the same food source, shelter or other resource.

CONDENSATION

The process by which matter changes from a gas to a liquid, e.g. when water vapour in the air becomes liquid water. Condensation is how clouds are formed.

CONSERVATION

The act of protecting species from extinction, protecting and enhancing ecosystems, and preserving biodiversity. With more than 42,100 species currently threatened with extinction worldwide, there is a growing global need for governments to invest in conservation programs.

The ringtail possum (top left) is known to engage in COPROPHAGY.

Fossiled dinosaur poo (left) is a type of COPROLITE.

Victoria's riflebird (*Ptiloris victoriae,* above) lives in north-eastern Queensland. Males have iridescent feathers, which they show off in a flamboyant, wing-swirling courtship display to attract females.

CONSUMER

An organism that feeds on living or dead organic material. See also *producer.*

CONVERGENT EVOLUTION

When unrelated species evolve similar characteristics.

The southern marsupial mole (*Notoryctes typhlops*) provides a great example of convergent evolution. It lives in the deserts of Australia and spends almost no time above ground, instead living underground like the European mole (*Talpa europaea*); although these animals have similar lifestyles and features, they are not closely related and evolved entirely independently.

COPROLITE

Fossilised faeces. Coprolites can tell scientists a lot about ancient diets.

COPROPHAGY

The act of eating faeces. Some Australian animals, including the **ringtail possum** (*Pseudocheirus peregrinus,* **above left**), eat large amounts of their poo to help absorb extra nutrients and fluid.

COPULATE

Sexual intercourse, mating.

COURTSHIP

Any behaviour animals use to attract or bond with the opposite sex, with the goal of mating.

CREPUSCULAR

An animal that is most active at dawn and dusk. See also *diurnal, nocturnal.*

CRUSTACEAN

A group of mostly aquatic arthropods that includes crabs, shrimp, barnacles and fish lice.

CUMULUS CLOUD

Fluffy looking clouds that have a flat base.

D

DECIDUOUS

Plants that drop their leaves each year, usually in autumn. Plants do this to survive harsh conditions during winter or in periods of low rainfall.

Most Australian trees are evergreen, but a few deciduous species grow here, including the **boab** (*Adansonia gregorii*, pictured) in northern Australia, and the deciduous beech (*Nothofagus gunnii*) of Tasmania.

The smooth knob-tailed gecko (above) uses DEIMATIC BEHAVIOUR to deter predators.

DECOMPOSER

An organism that breaks down organic matter into a usable form of nutrients. Decomposers are usually either bacteria or fungi.

The frill-necked lizard (left) in a stunning display of DEIMATIC BEHAVIOUR.

DEIMATIC BEHAVIOUR

Behaviour that is meant to intimidate and scare off predators; it is also known as a startle display.

The **smooth knob-tailed gecko** (*Nephrurus levis,* above) lives in arid areas of central Australia. When threatened, it raises its body, lunges and barks to intimidate predators. One of the most flamboyant examples of deimatic behaviour in Australia is that of the **frill-necked lizard** (*Chlamydosaurus kingii,* below left), which expands a colourful frill on its neck, somewhat like an umbrella.

DENDROCHRONOLOGY

The study of tree rings, which can reveal a tree's age and what the climate was like as it grew.

The pink dragon millipede from Thailand is a DETRITIVORE.

DEOXYRIBONUCLEIC ACID (DNA)
The genetic coding material of organisms and the building blocks of life.

DERMATOGLYPHICS
The study of fingerprints and skin patterns. Humans and other primates have fingerprints, and each individual's are unique. The only non-primates to have fingerprints are koalas and theirs can be so similar to a human's fingerprints that it can be difficult to distinguish between them.

DETRITIVORE
Organisms that feed on dead plant matter.

DETRITUS
Litter made up of dead material such as leaves, faeces and feathers.

DIAPAUSE
A temporary pause in growth and development that is typical of many insects and mites, which allows them to persist in seasonally variable environments. See also *embryonic diapause*.

DENDROCHRONOLOGY is the study of tree rings (above) to determine age.

Stink bugs (right) undergo periods of DIAPAUSE.

DICHROMATIC

When a species has two distinct colour morphs that are not due to an individual's age or breeding cycle.

The **eastern quoll** (*Dasyurus viverrinus,* top right) is dichromatic. Individuals of both sexes and all ages are born with either a pale fawn coat with white spots or a black coat with white spots.

Eastern quolls are DICHROMATIC.

DIMORPHISM

Physical differences that separate a species into two groups, such as males being much larger than females.

The male **eclectus parrot** (*Eclectus roratus,* above left) is mostly green and red, whereas the female (above right) is blue and crimson – a phenomenon also known as sexual dichromatism.

The numbat is DIURNAL.

Many land snails are DIOECIOUS.

DIOECIOUS
Used to describe plants and invertebrates in which the male and female reproductive organs are in separate individuals. See also *monoecious.*

DISPERSAL
When a young animal leaves its nest or parents to find its own home. Dispersal can result in high death rates, especially where there is intense competition for living spaces. Young male koalas are at high risk of being struck by cars or killed by dogs when they disperse from their natal range.

The **wedge-tailed eagle** (*Aquila audax,* right) can be found all across Australia. It has a wingspan of 2.3 m and builds nests weighing up to 400 kg. Young birds travel large distances when they disperse, sometimes more than 850 km.

DISTRIBUTION
The geographic area in which a species or a group of species occurs.

DIURNAL
Organisms that are most active during the day. The **numbat** (*Myrmecobius fasciatus,* top) is a diurnal marsupial that spends its days feeding on termites. See also *crepuscular, nocturnal.*

Wedge-tailed eagles fly great distances when they DISPERSE.

E

The platypus (above) uses ELECTROLOCATION to find prey.

ECHINODERM

A group of spiny marine invertebrates including starfish, sea cucumbers and sea urchins.

ECHOLOCATION

A way to detect obstacles and find prey by listening for sounds reflected by objects. The white-striped freetail bat (*Austronomus australis*) is one of more than 60 species of microbat in Australia and is one of the few species whose echolocation calls can be heard by humans. Most bat species make sounds at such high frequencies that they are out of our hearing range.

Starfish (above) are ECHINDODERMS.

ECOLOGY

The study of the relationships between different organisms and their environment.

ECOSYSTEM

A way to describe a stable system created by communities of interacting organisms and their physical environment.

ECOTONE

The transition between two different communities. For example, where two different types of forest meet and blend into each other.

ECTOTHERM

Animals that regulate their body temperature using different behavioural tactics, such as basking in the sun or lying in the shade. Reptiles and invertebrates are ectotherms. See also *endotherm*.

The saltwater crocodile is an ECTOTHERM.

ELAPHINE

Related to, or resembling, red deer (*Cervus elaphus*). Deer were introduced to Australia in the 1860s by *acclimatisation societies*.

ELECTROLOCATION

Detecting objects by sensing the electrical field they give off.

The **platypus** (*Ornithorhynchus anatinus*, top left) closes its eyes, nose and ears when it dives underwater to search for food. It uses special electroreceptors in its bill to detect electrical signals produced by muscle movement in its prey.

EMBRYO

The first stage of development in multicellular organisms.

EMBRYONIC DIAPAUSE

A temporary pause in growth and development. The **red kangaroo** (*Osphranter rufus*, right) is found in desert areas of Australia. It is not only the continent's largest kangaroo but is also the biggest native land mammal and the largest living marsupial. Females can pause pregnancy, stopping the development of an embryo and then restarting it when conditions are more favourable. The much smaller tammar wallaby (*Notamacropus eugenii*) undergoes the longest periods of embryonic diapause, putting its embryos 'on hold' for almost a year!

The female red kangaroo uses EMBRYONIC DIAPAUSE when needed.

Leadbeater's possum (left) is critically endangered.

ENDANGERED

A category of threat assigned to organisms that have a high risk of becoming extinct in the near future. According to the International Union for Conservation of Nature (IUCN), threatened species are classified as critically endangered (CE), endangered (EN) or vulnerable (VU). More than 42,000 species around the world are at risk of becoming extinct. At least 16,000 of them are listed as endangered, and more than 9000 are critically endangered. Unfortunately, Australia has many critically endangered plants and animals, such as **Leadbeater's possum** (*Gymnobelideus leadbeateri*, above), the faunal emblem of Victoria.

ENDEMIC

Organisms that only occur in a particular country or area. In Australia, 87% of mammals, 93% of reptiles and 92% of vascular plants occur nowhere else in the world!

ENDOSKELETON

The internal skeleton that exists within the body of vertebrates. See also *exoskeleton*.

ENDOTHERM

Animals that regulate their body temperature by using internal mechanisms such as changing the size of blood vessels, sweating, panting and shivering. Birds and mammals are examples of endotherms. See also *ectotherm*.

ENTOMOPHILOUS

Plants that are pollinated by insects. The Tuncurry midge orchid (*Genoplesium littorale*) is a critically endangered species that is pollinated exclusively by chloropid flies.

ENVENOMATION

A type of poisoning that results from the bite, sting or injection of venom from an animal or plant.

Australia has some of the most venomous animals in the world, including the **inland taipan** (*Oxyuranus microlepidotus*, right), the geographer cone snail (*Conus geographus*) and the box jellyfish (*Chironex fleckeri*).

Bird's nest ferns (above) are EPIPHYTES.

Bearded dragons are ENDEMIC to Australia.

Native midge orchids (left) are ENTOMOPHILOUS.

ENVIRONMENT

The place where an organism lives, or the natural world, or a particular area on Earth.

ENVIRONMENTAL HETEROGENEITY

Differences in plant species, plant structure and abiotic components, such as soil and climate, across an area. Heterogeneity promotes a greater diversity of species by providing different habitats and resources for species to use within an area.

EPICORMIC GROWTH

Shoots that grow from the bark, stem, trunk or branch of a plant in response to damage or stress. This is a common response after eucalypts have been burnt or have had limbs removed.

EPIPHYTE

Plants that grow on other plants or objects purely for physical support.

The **bird's nest fern** (*Asplenium australasicum*, top right) grows in wet forest and rainforest in eastern New South Wales and Queensland. These ferns typically flourish in the forks of rainforest trees, although they can also grow on rocks or in suitable soil, and can reach up to 1.5 m wide.

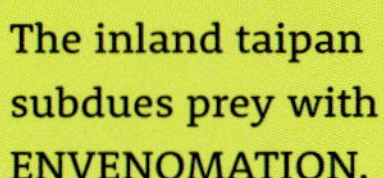

The inland taipan subdues prey with ENVENOMATION.

The short-beaked echidna (above) is EURYECIOUS.

EROSION

A geological process where materials are worn away and moved elsewhere by forces such as wind or water. Erosion has led to a lot of the famous landmarks we see today in Australia, which were once part of larger landmasses, including the monolith of Uluṟu.

ESTUARY

Bodies of brackish water and wetlands found where rivers and creeks flow into the sea.

EUKARYOTE

Organisms that have a complex cell structure, with cells that have a nucleus bound in a membrane. See also *prokaryote.*

EURYECIOUS

Organisms that can live in a variety of different habitats. The **short-beaked echidna** (*Tachyglossus aculeatus,* left) is Australia's most widely distributed mammal. It can survive in most habitats, whether in freezing Tasmania or arid Western Australia and the Northern Territory.

EUSOCIALITY

Species that live in colonies where only one female produces offspring and the other individuals attend to her needs, e.g. bees and ants.

EUTROPHICATION

The accumulation of different nutrients in a body of water such as a lake or river. This high level of nutrients can lead to poor water quality and a lack of oxygen in the water, resulting in dead zones where aquatic animals cannot survive.

The Remarkable Rocks on Kangaroo Island, off SA, are abstract marvels of natural EROSION.

The Christmas Island red crab has an EXOSKELETON.

The thylacine (above) is EXTINCT.

EVAPORATION

When liquid water changes into water vapour and enters the atmosphere. This happens when heat breaks apart the bonds that hold water molecules together and it is why puddles disappear so quickly on a hot day.

EVOLUTION

Changes to the characteristics and diversity of organisms over time. See also *natural selection*.

EXOSKELETON

The rigid, protective shell or covering that encases the body of an arthropod such as the **Christmas Island red crab** (*Gecarcoidea natalis,* above). See also *endoskeleton*.

EXOTIC

An organism that does not naturally occur in an area but has been introduced from elsewhere.

EXTANT

Species that currently exist on Earth. See also *extinct*.

EXTINCT

Species that once existed on Earth but no longer have any living individuals. This term is also applied to organisms that are locally extinct from specific geographic areas where they once occurred. This can be a natural process that happens with long-term environmental changes or may happen as a result of human activities. See also *extant* and *extirpated*.

Australia has the worst mammal extinction rate of any country in the world. Since European colonisation, at least 39 mammals have become extinct. These include the well-known Tasmanian tiger, also known as the thylacine, (*Thylacinus cynocephalus,* top) and some lesser-known species such as the desert rat-kangaroo (*Caloprymnus campestris*), the Christmas Island pipistrelle (*Pipistrellus murrayi*), the Lake Pedder earthworm (*Hypolimnus pedderensis*), and the pig-footed bandicoot (*Chaeropus ecaudatus*).

EXTINCTION DEBT

Used to describe the loss of species that are projected to become extinct in the future due to events that have already occurred, such as habitat destruction or urbanisation.

EXTIRPATED

The localised eradication or extinction of a species from a part of its range.

EYESPOT

A spot on part of an animal's body that resembles an eye and may act defensively to trick predators. See also *ocellus*.

F

The bare-nosed wombat is a FOSSORIAL mammal with cube-shaped FAECES.

FAECES

Faeces are waste material that is expelled through the anus of an organism (also known as poo).

The **bare-nosed wombat** (*Vombatus ursinus,* top) has cube-shaped faeces, but not because it has a cube-shaped anus! Its faeces are square because they take 14–18 days to travel through its digestive system, spending most of that time in the marsupial's large intestine, which is lined with horizontal ridges.

FAUNA

Another word for animals.

FECUND

Extremely fertile or able to produce a large number of offspring in a short space of time.

FISH

Legless aquatic vertebrates that use gills to breathe.

The largest living fish is the **whale shark** (*Rhincodon typus,* right), which can be found at certain times of the year off the Western Australian coast. These endangered fish grow to more than 18 m in length.

FITNESS

How well adapted an individual animal is to its environment.

FLORA

Another word for plants.

FOLIVORE

Animals that eat leaves. See also *carnivore, frugivore, herbivore, omnivore*.

The **koala** (*Phascolarctos cinereus,* right) feeds mainly on leaves from eucalyptus trees.

Some FOSSILS reveal the earliest known birds, such as *Archaeopteryx*.

FOSSIL

The preserved remains of an organism that lived thousands of years ago, including bones, footprints and impressions. Fossils are usually preserved in rock, and in Australia, some are even opalised.

The oldest fossils ever found are from Archaean rocks in Western Australia. These fossils reveal that micro-organisms existed almost 3.5 billion years ago.

Flying foxes (above) and wompoo fruit doves (below) are FRUGIVORES.

FOSSORIAL

Animals that burrow.

FRUGIVORE

Animals that primarily eat fruit, e.g. flying foxes (*Pteropus* spp.) or **wompoo fruit doves** (*Megaloprepia magnifica*, above). See also *carnivore*, *folivore*, *herbivore*, *omnivore*.

FUNGI

Fungi are eukaryotic organisms that have their own kingdom separate from plants and animals. They include yeasts, rusts, smuts (a fungal plant disease), mildews and mushrooms.

Australia has some 15,000 known recorded species, but it has been estimated that only 5% of Australia's total fungi have yet been discovered. Some species are visible to the naked eye, but others are microscopic.

FUNICULUS

In plants, the stalk or thin cord that attaches a seed to a plant or fruit, providing water or nutrients (similar to the umbilical cord of placental mammals). When the seed matures, dries out and is ready to grow, the funiculus attachment is severed.

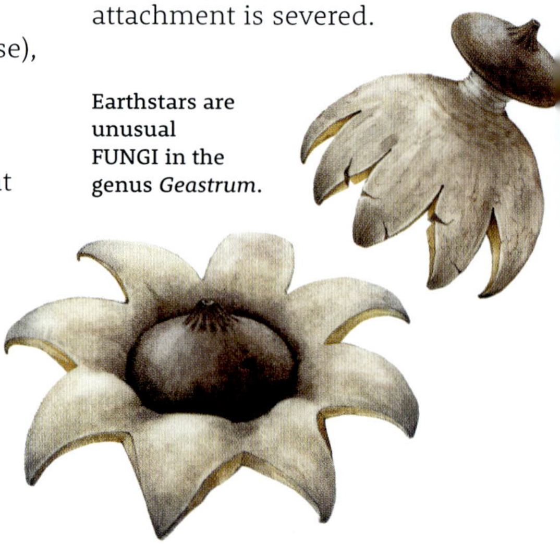

Earthstars are unusual FUNGI in the genus *Geastrum*.

G

GALAXY

A group of billions of stars, their solar systems, gases and dust, all held together by gravity.

GENETIC ENGINEERING may one day help control feral cats.

GASTROLITH

A stone an animal has swallowed which sits in the stomach or gizzard and aids in digestion and breaking up food. The emu (*Dromaius novaehollandiae*) and the **southern cassowary** (*Casuarius casuarius,* above) both eat gastroliths to help grind up food.

GENE

A gene is a segment of DNA and is how heritable traits are passed on from a parent to its offspring.

GENETIC DRIFT

Random changes in the frequency of genetic traits within a population.

GENETIC ENGINEERING

Using technology to alter an organism's genes.

In Australia, **feral cats** (*Felis catus*, above) kill more than three billion native animals each year. Gene editing is being explored to reduce the number of feral cats in a bid to save native species. Scientists are considering altering genes in released female cats so that they produce only male offspring. Over time, this trait would hinder reproduction in the feral population and reduce the number of these predators.

GENOTYPE

The genetic make-up of an organism. Genotype usually refers to a particular set of

Fox Glacier/ Te Moeka o Tuawe is a 13-km long GLACIER on New Zealand's South Island.

The leaf sheep nudibranch has GILLS outside of its body.

genes an organism has inherited. See also *phenotype*.

GENUS
A rank used in Linnaean taxonomy that is below family but above species and is italicised, e.g. the **spotted fritillary** (left) of Central America is in the genus *Melitaea*.

GEOLOGY
The study of the rocks that make up Earth and the many processes that have created and changed these rocks over time.

GERMINATION
When a plant seed sprouts and begins to grow.

GESTATION
The development of an embryo inside its parent before it is born.

GILLS
Organs that aquatic animals use to breathe. Water flows past the gills, which allows oxygen to be absorbed into the body. The word nudibranch means 'naked gill' because the gills of these marine invertebrates are on the outside of the body.

GLACIER
A large body of compacted ice, snow, rock and sediment that moves slowly over land.

GLAND
An organ or group of cells that influence the endocrine or exocrine system by making and releasing a substance into a duct or into the blood.

GEOLOGY is the study of rocks, stones and gems.

GONDWANA

The supercontinent that existed in the Southern Hemisphere before Africa, Antarctica, Australia, India and South America broke apart.

The Tasmanian pencil pine (below) is an ancient GYMNOSPERM.

GPS TRACKING

Attaching a satellite-trackable device to monitor an animal's movements. See also *radio tracking*.

GRAZER

A herbivore, such as a wombat, pademelon or horse, that feeds on low-growing plants and grasses. See also *browser*.

GREENHOUSE GAS

A gas that absorbs and emits radiant energy. These gases are effective at trapping heat in Earth's atmosphere. Some greenhouse gases are necessary – they keep the planet warm enough for us to survive – but human activities are increasing the levels of greenhouse gases in the Earth's atmosphere, which can result in global warming.

GROOMING

Behaviours an animal engages in to care for the surface of its body. Mutual grooming, or allogrooming, is when animals groom each other, which strengthens social bonds between individuals.

GYMNOSPERM

Woody plants that don't flower but produce seeds on the surface of their reproductive structure, rather than inside a fruit. The **pencil pine** (*Athrotaxis cupressoides,* above left) is a gymnosperm from Tasmania that has existed since Gondwanan times. A single pencil pine can live for more than 1200 years.

Zebras are GRAZERS that engage in mutual GROOMING.

HABITAT
The environment where an organism or a species usually lives.

Sheep are HERBIVORES that live in HERDS.

The eastern montane green cockroach is HEMIMETABOLOUS.

HABITUATION
When an organism is so frequently exposed to a stimulus that their response reduces over time. For example, animals that see people a lot are often less afraid of people than animals that rarely encounter humans.

HADAL ZONE
The deepest part of the ocean, which lies in trenches below the abyssal zone, from 6000–11,000 m.

HAEMAL
In anatomy, an adjective relating to the part of the body that contains the heart.

HELIACAL
An adjective meaning of or near the sun.

HELOTISM
When an organism enslaves another for its own benefit. Such a relationship exists between the dominant fungus and the enslaved algae that together make up lichen.

HEMIMETABOLOUS
An insect that transitions directly from larva to adult without a pupal stage, e.g. cockroaches, grasshoppers and dragonflies.

HEMIPENES
Paired internally concealed penises found in some reptiles.

HEMIPTERA
An order of insects, or bugs, with piercing mouthparts, e.g. cicadas and aphids.

HERITABLE
A characteristic that is capable of being passed on to offspring.

HERBIVORE
An animal that eats only plants as a food source, e.g. koalas, rabbits or sheep. See also *carnivore, folivore, frugivore, omnivore*.

HERD
A large group of animals that live together.

HETEROECIOUS
An adjective to describe a parasite that uses different host organisms at different stages of its life cycle. See also *autoecious*.

HETEROTROPH

Organisms that cannot make their own food and must consume other living or dead organisms to survive. Carnivores, detritivores, herbivores and omnivores are all heterotrophs. See also *autotroph, chemosynthesis, photosynthesis*.

The carnivorous **Tasmanian devil** (*Sarcophilus harrisii,* above) is a scavenger that eats live prey as well as cleaning up the carcasses of dead animals within its habitat, which is helped by its incredibly strong bone-crunching jaws.

HIBERNATION

A state animals enter to survive winter, during which their body temperature and metabolism can drop to near zero.

In Australia, it's not cold enough for snakes to hibernate (they brumate instead), but four species of pygmy-possum, a handful of bats and the short-beaked echidna are all known to hibernate for extended periods. The eastern pygmy-possum *(Cercartetus nanus)* – found along the east coast from Queensland to South Australia and throughout Tasmania – holds the worldwide title for the longest hibernation of any mammal, spending an entire year in hibernation. See also *aestivation, brumation, torpor*.

The eastern pygmy-possum HIBERNATES in winter.

Jellyfish occupy the hadal zone.

Turkey tail fungus (top) is made from HYPHAE, which form underground networks of mycelia (below).

Lotus leaves (above) are HYDROPHOBIC.

Magpies (left) get territorial over their HOME RANGE during nesting season.

HOME RANGE
The area or territory where an animal usually lives. Some animals use scent markings to mark the edges of their range, and some aggressively defend the area from others.

HOMEOSTASIS
When organisms or particular elements of ecosystems work together to achieve balance or equilibrium.

HUMIDITY
A measurement of water vapour in the air.

HUMUS
Partially decomposed organic matter in soil that provides nutrients for plants and increases how much water the soil can absorb.

HYDROLOGY
The study of where water occurs, how and where it moves, and how to manage it. Australia is the world's driest inhabited continent, so hydrology is very important.

HYDROPHILIC
Molecules and surfaces that can become wet and absorb water. See also *hydrophobic*.

HYDROPHOBIC
Molecules and surfaces that are resistant to absorbing water. See also *hydrophilic*.

HYPHAE
Filamentous, elongated strands of fungi that grow beneath the soil and absorb nutrients, combining to make an underground network called mycelium.

HYPOTHESIS
A proposed explanation that can be tested by experimentation.

I

IDIOTHETIC

When an animal uses information from its previous orientation to navigate without the use of spatial cues from sight, sound or smell. The word comes from a Greek word meaning 'self-proposition'. Path integration is a type of idiothetic navigation known to be used by bees (pictured), rats and crabs. See also *allothetic*.

Both the male and female brolga INCUBATE their eggs.

INBREEDING DEPRESSION

When animals breed with close relatives and lose genetic variation, causing their growth, survival, or capacity to breed to be reduced. This may happen to animal populations that are cut off from other communities by habitat destruction or geological barriers.

Animals that live on islands are particularly at risk of inbreeding depression, such as the Barrow Island population of the **black-flanked rock-wallaby** (*Petrogale lateralis lateralis*, right). See also *outbreeding depression*.

INCUBATE

Sitting on an egg to keep it warm until it hatches. This term is usually applied to birds, such as swans, **brolgas** (*Grus rubicunda*, above), ducks and hens, but the egg-laying platypus (*Ornithorhynchus anatinus*) also incubates its eggs.

Black-flanked rock-wallabies are at risk of INBREEDING DEPRESSION.

IMPRINTING

When animals immediately after birth or hatching focus on one thing (almost always their mother) to form an attachment that helps them learn how to survive by copying everything she does. Introduced **Mallard ducklings** (*Anas platyrhynchos*, above) and black swan cygnets (*Cygnus atratus*) imprint after hatching. Occasionally, an animal will imprint on a human caretaker or another animal.

The green tree frog (right) is an INDICATOR SPECIES.

The northern sword-grass brown butterfly (below) is an INSECT.

INDICATOR SPECIES

A species that exists only under certain environmental conditions so is used to determine the health of an ecosystem.

Frogs, such as the **green tree frog** (*Litoria caerulea,* above), have permeable skin, which they breathe through and which water can pass through. This makes them particularly sensitive and vulnerable to pollutants. They are used as an indicator of water quality and ecosystem health.

INNATE BEHAVIOUR

Behaviour that is not learned but passed on genetically, including the drive to suckle, which is inherent in mammalian young.

INORGANIC

Substances that are not made from living material, e.g. rocks and metals.

INSECTS

Arthropods with an exoskeleton, three pairs of legs and one pair of antennae. This group includes ants, bees, beetles, butterflies, flies and wasps.

There are more insects in the world than any other group of organisms. Australia alone is estimated to have more than 200,000 insect species.

Suckling is INNATE BEHAVIOUR for mammals.

Camels are an INTRODUCED SPECIES in Australia.

Dingoes (above) are ITEROPAROUS.

Rainbow stag beetles (below) are IRIDESCENT.

INTRODUCED SPECIES

Also known as invasive species, these are animals and plants that have been purposefully or accidentally moved out of their natural range by humans.

Introduced species in Australia include **camels** (*Camelus dromedarius,* above), rabbits (*Oryctolagus cuniculus*), and plants such as lantana (*Lantana camara*) and prickly pear (*Opuntia spp.*). Introduced species have contributed to the extinction of at least 79 native Australian species.

INVERTEBRATES

Animals that do not have a backbone. This large group includes insects, crustaceans, and molluscs.

IRIDESCENT

Surfaces that appear to change colour depending on which angle you look at them. Many beetles, like scarabs, have iridescent shells, and some birds have iridescent feathers, such as the glossy ibis.

More than 85 species of **peacock jumping spider** (*Maratus spp.*) live in Australia. Males raise their iridescent, vividly coloured abdomens and dance to attract females, in an elaborate courtship display.

ITEROPAROUS

Capable of mating and reproducing more than once in a lifetime. Placental mammals, like humans and dingoes, are typically iteroparous. See also *semelparous*.

JACENT
Lying flat, or sluggish.

A perentie (monitor lizard) lying JACENT in warm sand.

A JASPERATED landscape at Marble Bar, WA.

The semi-precious stone jasper can be green or a JACINTHE orange colour.

JACINTHE
A moderate orange colour.

JACTATION
A tossing, twitching or jerking of the body; it is a characteristic of severe fevers.

JASPERATED
Mottled or streaked with various colours, like the semi-precious stone jasper.

JERBOA
A small, nocturnal desert-dwelling rodent from north Africa and Asia. The Australian **kultarr** (*Antechinomys laniger,* right) was referred to as a 'jerboa' by early European naturalists; however, it is not a rodent but a carnivorous marsupial in the order Dasyuromorphia.

The kultarr was called a JERBOA by early European naturalists.

Despite its name, the lion's mane jellyfish isn't truly JUBATE.

Salt marsh rush is JUNCACEOUS.

JUBATE

Having or possessing a mane. The most famous maned animal is arguably the lion, but the **lion's mane jellyfish** (*Cyanea capillata,* above) is so-named for the 'mane' of long, hair-like tentacles (over 1200!) that hang from its bell.

JUGATE

Side by side, or overlapping in pairs, such as a pinnate leaf.

JUGULAR

The neck or throat area of an animal, including humans. Africa's big cats are known for killing their prey quickly by 'going for the jugular'.

JUNCACEOUS

Of, like, or pertaining to rushes.

Austral rush (*Juncus australis*) features a tight cluster of flowers and is native to south-eastern Australia. **Salt marsh rush** (*Juncus kraussii,* right) occurs in all states of Australia.

JURASSIC

An era that describes the time period from about 200 million years ago to 145 million years ago. This is an important period during which dinosaurs flourished and from which the first fossils of birds were found.

JUVENAL

The plumage stage of an altricial bird, which is underdeveloped when it hatches and requires care before it is ready to leave the nest. See also *altricial.*

K

KARST

A type of landscape built on a base of limestone, which erodes and dissolves over time to create fissures, stacks, sinkholes, caves and other interesting landforms.

Australia has some of the most spectacular karst landscapes in the world, including the **Pinnacles** in Nambung National Park, WA (pictured), and the coastline along the Great Australian Bight, SA.

KERATIN

A protein that is used in creating hard structures on animals, such as horns, nails, scales and hair.

KEYSTONE SPECIES

A species that is important to the functioning of the entire ecosystem and helps define it. If this species is removed, other species will not be able to survive.

The **southern cassowary** (*Casuarius casuarius*, right) from north Queensland is a keystone species. It is important for the dispersal of plants, transporting the seeds of more than 60 species in its faeces and on its feet as it moves through the rainforest.

Gilbert's potoroo (*Potorous gilbertii*) is a keystone species found in the South West of Western Australia. Its diet depends on 40 kinds of fungi, which need the potoroo to disperse their spores and in turn provide the potoroo with unique nutritional benefits.

A rough-scaled python's scales are made of KERATIN.

The superb lyrebird (below) is considered a K-SELECTION species.

KINESICS

The study of nonverbal communication or behaviour, such as the use of gestures, postures or facial expressions to communicate. In animals, most studies have focused on the kinesics of domesticated companion animals, such as dogs and cats, or on primates, which often use similar facial expressions and postures as humans.

KINESICS help us better understand a kelpie's expressions.

KINGDOM

The second-highest ranking in Linnaean taxonomic structure.

KRONISM

A type of cannibalism where parents eat their offspring. Nutrient-deficient hamster mothers will sometimes consume their newborns. Male blenny (*Ecsenius* sp.) fish have also been known to consume all the eggs of their offspring, particularly if it is only a small brood, to enable them to breed with a new female and achieve a larger brood.

K-SELECTION

K-selected species produce a smaller number of offspring and provide a higher level of parental care. Their offspring have greater rates of survival, but this requires a high level of investment from the parent. See also *R-selection*.

The **superb lyrebird** (*Menura novaehollandiae,* above) uses this strategy. Females typically only lay one egg in a year. The mother is very dedicated to defending the nest from predators, and the chick will stay with the mother until the next breeding season.

LAVA snakes over the Earth in Iceland.

L

LARVAE

A life cycle-stage in animals after they have hatched from an egg and before they change into their adult form. This term is applied to fish, amphibians and insects.

LAVA

Molten liquid rock on the surface of Earth. Lava comes out of a volcano or a crack in the Earth's crust. Lava is very hot – it can be more than 1000 °C. See also *magma*.

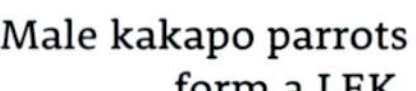

Male kakapo parrots form a LEK.

Black soldier flies spend part of their lives as LARVAE.

LEARNED BEHAVIOUR

Animal behaviour that is not genetically programmed but learned over time through different experiences.

LEK

A breeding strategy where males of a species gather in one area to perform courtship displays. Females visit the group and compare the males' performances to decide which has the most impressive display.

The critically endangered owl parrot or **kakapo** *(Strigops habroptila,* above*)* is endemic to New Zealand and forms a lek. Each male digs a shallow bowl – together creating a communal area that can stretch more than 200 m. The male then puffs up his chest and starts his display, accompanying it with loud booming or honking noises approximately every two seconds. Peacocks, manakins and some birds of paradise exhibit similar courtship rituals in leks.

LIGHTNING crackles over Kata–Tjuṯa, NT.

A LEUCISTIC sugar glider.

LIGHT POLLUTION disorientates green turtle hatchlings.

LEUCISM

A genetic mutation in which animals lack pigment in some or all of their feathers or fur but still have some typical pigment in their skin and eyes. See also *albinism*.

When a **sugar glider** (*Petaurus breviceps,* right) is leucistic, it has white fur but retains dark colour in its eyes.

LIFE CYCLE

Certain developmental changes an organism undergoes during the course of its life.

LIGHTNING

A huge spark of electricity between clouds and the ground. One bolt of lightning can produce 500 million volts and a temperature of about 27,000 °C. That's three times hotter than the surface of the Sun!

During summer, the Australian continent experiences about five major storms every day, each producing up to 3000 cloud-to-ground lightning strikes per hour.

LIGHT POLLUTION

Light pollution is the presence of artificial light, which has negative effects on species and can alter important biorhythms. Artificial light can disrupt sleep cycles, disorient migratory birds and sea turtles, and reduce hunting time for predators and the availability of prey.

The bacteria that turn Lake Hillier, WA, pink show high LIMITS OF TOLERANCE for salinity.

The Wollemi pine is a LIVING FOSSIL.

A LUNAR ECLIPSE is best viewed with a telescope.

LIMITS OF TOLERANCE

The upper and lower limits of environmental factors – such as temperature, light or salinity – that an organism can tolerate. Some organisms have very broad limits of tolerance and some very narrow.

Some archaea and alga, such as *Dunaliella salina*, can tolerate the high salinity environments of Australia's salt lakes and turn them bright pink in the process.

LIVING FOSSIL

An extant species that has existed for a long time unnoticed and resembles extinct fossil relatives.

The **Wollemi pine** (*Wollemia nobilis,* top left) is an Australian living fossil. These trees were known only from fossils – some dated to 200 million years old – until 1994, when living trees were discovered in a remote gorge in Wollemi National Park, north-west of Sydney.

LUNAR ECLIPSE

When the Earth moves between the Sun and the Moon, the planet's shadow temporarily obscures the Moon, making the moon glow red and dark. A lunar eclipse only happens at night during a full moon. Such eclipses happen twice to five times a year, although five are rare. See also *solar eclipse.*

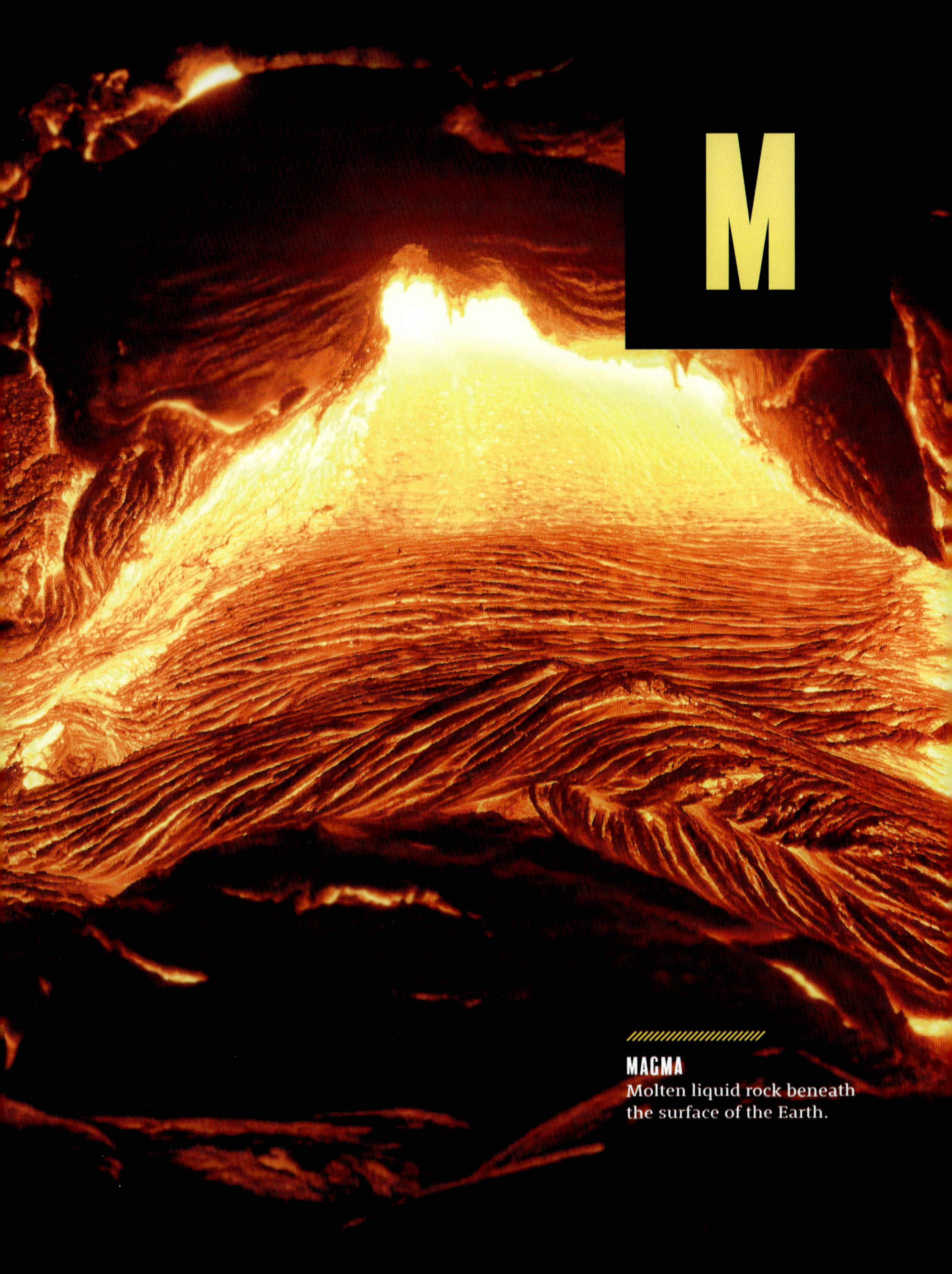

M

MAGMA
Molten liquid rock beneath the surface of the Earth.

MALACOLOGY is the study of molluscs like the flamboyant cuttlefish (left).

MEGAFAUNA, such as the giant lizard *Megalania* (below left) and thunder birds like *Genyoris* (below right), used to inhabit Australia but are now extinct.

MALACOLOGY

The study of molluscs, which includes cuttlefish, squid, octopuses, clams, mussels, snails, slugs, and sea slugs (nudibranchs). See also *mollusc*.

MAMMALS

Endothermic vertebrates that mostly have hair (except for cetacean marine mammals) and produce milk from mammary glands to feed their young.

MARK-RECAPTURE TECHNIQUE

A technique used to estimate the size of an animal population. Individuals are caught, marked in some way, and then released. Trapping is usually done again at a future date, and the proportion of marked individuals caught versus new individuals can be used to estimate the overall size of the population.

MARSUPIALS

A group of non-placental mammals that have short gestation times. They give birth to live young that are not very developed, so their offspring spend a long period of time in their mother's pouch to develop further.

Worldwide, there are more than 330 marsupial species, and two-thirds of them live in Australia, including herbivorous koalas, greater gliders, wombats, wallabies and kangaroos; omnivorous bandicoots, possums, pygmy-possums and potoroos; and carnivorous small dasyurids, the numbat, quolls and the Tasmanian devil.

MEGAFAUNA

Animals weighing more than 40 kg, such as elephants and rhinoceroses. In Australia, this term is used to describe gigantic animals that once dominated Australian landscapes until they became extinct 45,000 to 7000 years ago.

Australia's megafauna included 450-kg kangaroos; 2000-kg wombats like *Diprotodon*; 7-m-long lizards such as *Megalania*; 180-kg flightless 'thunder birds'; 130-kg marsupial lions like *Thylacoleo*, and even car-sized tortoises with clubbed tails!

MEGAPODES

Stocky, mound-building ground birds, such as the malleefowl (*Leipoa ocellata*), orange-footed scrubfowl (*Megapodius reinwardt*) and the **Australian brush-turkey** (*Alectura lathami,* below).

MEIOSIS

Cell division where a single cell divides twice, resulting in four cells with half the genetic information. See also *mitosis.*

MELANIN

A brown to black pigment that colours the iris, hair and skin of animals and humans – in which it increases with sun exposure (i.e. tanning).

The honey possum is a tiny MARSUPIAL that is endemic to Western Australia.

The Australian brush-turkey is a mound-building MEGAPODE.

The viceroy butterfly (above) MIMICS the monarch butterfly (right).

METAMORPHOSIS of a monarch butterfly (above).

MELANOPHORE

A type of skin pigment cell, or chromatophore that makes and stores melanin. Melanophores are known as melanocytes in humans.

MELLITOCHORY

A method of seed dispersal where plant seeds are carried to new places by bees (below).

METAMORPHOSIS

When an organism changes from an immature, often larval, form to an adult form. This phenomenon is most pronounced in insects such as butterflies and in amphibians that metamorphose from tadpoles to frogs.

METEOR

Rocks from space, such as bits of an asteroid, which enter Earth's atmosphere at high speed and burn up. See also *asteroid*.

MICROBIOLOGY

The study of organisms that are so small they are invisible to the human eye.

MICRO-ORGANISM

An organism that can only be seen through a microscope.

MIGRATION

When animals move from one area to another at certain times of the year and then back again.

Some female **green turtles** (*Chelonia mydas,* below right) travel more than 2600 km in their migration between feeding grounds and nesting beaches. Those that nest on the beaches of the Great Barrier Reef, in Queensland, migrate from feeding grounds in Indonesia, Papua New Guinea, Vanuatu, New Caledonia and across northern Australia.

Rainbow lorikeets are known to engage in MOBBING other birds they compete with for food.

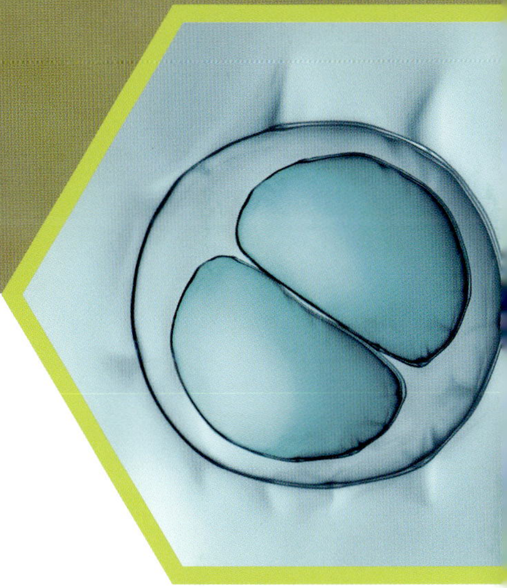

MIMICRY

When an animal copies the behaviour, appearance, scent or call of another species. Many Australian birds mimic other bird calls.

The superb lyrebird (*Menura novaehollandiae*) is one of the most famous. In their vibrant courtship displays to attract females, the males mimic the calls of more than 20 other birds. Although the males usually get all the attention for their singing, female lyrebirds can also replicate a range of other birds' calls.

MITOSIS

Cell division where a single cell divides into two cells with identical genetic information. See also *meiosis*.

MOBBING

When a group of animals works together to harass a predator or competitor. **Rainbow lorikeets** (*Trichoglossus moluccanus,* above) are well-known 'mobsters' that gang up to harass other birds.

MOLECULE

The smallest identifiable unit of a particular substance.

MOLLUSC

Invertebrates that usually have an unsegmented, soft body and a muscular foot or tentacles, as well as an internal or external shell and a radula. Molluscs include mussels, octopuses, sea slugs, nautiluses, snails and squid.

MITOSIS (above) occurs when a single cell splits into two.

The paper nautilus (above) is a MOLLUSC.

Monitor lizards (left) have MONOCULAR VISION.

Kuni's nudibranch is MONOECIOUS.

The platypus (right) is a MONOTREME.

MONOCULAR VISION

When an animal has eyes on either side of its head and can see with one eye at a time, giving a wider field of vision, e.g. monitor lizards, rabbits and horses. See also *binocular vision*.

MONOECIOUS

An organism that has female and male sex organs occurring on the same individual. Individuals of **Kuni's nudibranch** (*Goniobranchus kuniei*, above right) each have both male and female reproductive organs, but they still need to find another individual to reproduce with. Being monoecious means they can find any individual from their species, not just one of the opposite sex. See also *dioecious*.

MONOGAMY

A relationship strategy whereby animals have only one sexual partner at a given time. Australia's **black swans** (*Cygnus atratus*, below) are highly monogamous, with about 94% of them mating for life. See also *polygamy*.

MONOTREME

Mammals that lay eggs rather than produce embryos that develop internally. Several species of echidna and the **platypus** (*Ornithorhynchus anatinus*, above) are the world's only surviving monotremes.

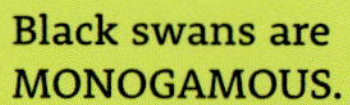

Black swans are MONOGAMOUS.

Peruvian poison dart frogs (*Ranitomeya* spp.) are look-alikes that employ MÜLLERIAN MIMICRY.

Slime moulds (below) form MULTICELLULAR ORGANISMS to reproduce.

MORPHOLOGY
The form, structure and anatomy of an organism.

MOTILE
Organisms that are able to independently move around. See also *sessile*.

MOULT
When an animal periodically sheds its feathers, hair or skin so that new feathers, hair or skin can grow.

MUCILAGE
A thick secretion from plant roots and seeds that contains polysaccharides and protein.

MÜLLERIAN MIMICRY
When species of plant or animal that are unpalatable or poisonous evolve to look like one another, discouraging predators from eating either species. See also *aposematic, Batesian mimicry.*

MULTICELLULAR ORGANISM
An organism made up of more than one cell. All animals and land plants are multicellular. See also *unicellular organism.*

Slime moulds are organisms that are neither in the plant nor animal kingdom, but in a kingdom called Protozoa. There are more than 260 species in Australia, and they live as unicellular organisms until it comes time to reproduce. The unicellular organisms then join to form a multicellular fruiting body to release spores.

MYCOLOGY is the study of fungi (right).

Clownfish have a MUTUALISTIC relationship with anemones.

Fairy lanterns (*Thismia* spp.) are MYCOHETEROTROPHS.

MUTATION

A change in a gene or chromosome during replication or reproduction. Mutations can cause a change in an organism's structure or behaviour and are the process behind evolution. Some mutations are beneficial, but others can be harmful.

MUTUALISTIC

A symbiotic relationship between two different species where both species benefit from the association. See also *commensalistic, mycorrhizae, parasitic, symbiotic*.

MYCOHETEROTROPH

A type of plant that gets part or all of its food from fungi, rather than from photosynthesis, e.g. *Dipodium* species of orchid.

The fairy lantern (*Thismia rodwayi*) is a small plant that grows on forest floors in New South Wales, Victoria and Tasmania. This fascinating plant is another mycoheterotroph. Instead of producing its own food through photosynthesis, it gets its food from its association with fungi.

MYCOLOGY

The study of fungi, including their taxonomy, genetic properties and use to humans.

MYCORRHIZAE

A symbiotic relationships between fungi and plant. The fungus colonises the host plant's roots in a sometimes-mutualistic association. See also *commensalistic, mutualistic, parasitic, symbiotic*.

N

NATIVE

A species that occurs in an area through natural processes, not through human intervention. The **tawny frogmouth** (*Podargus strigoides,* left) is native to the Australian mainland and Tasmania.

Malleefowl chicks are NIDIFUGOUS and are on their own as soon as they hatch.

A sea krait (right) uses NEUROTOXIN to defend itself from predators.

Studies have found that the bite of a white-tailed spider (below) doesn't actually cause NECROSIS.

NATURAL SELECTION

An evolutionary process whereby individual organisms that are best adapted to their environment survive and go on to reproduce. Organisms that are poorly adapted to their environment do not survive and do not go on to reproduce. See also *artificial selection, sexual selection*.

NECROSIS

The cell death of a piece of body tissue on an organism. This can happen from exposure to radiation, injury or disease.

White-tailed spiders (*Lampona spp.*, left) live in southern and eastern Australia. There was a widespread belief that their bite caused necrotic sores, but studies have shown this is not the case. However, a bite from a recluse spider, a species native to North and South America, can certainly cause necrosis.

NEONATE

A mammal from birth to four weeks of age. Many changes occur in the neonatal period.

NERITIC

Parts of the ocean that are roughly between the low tide level to 200 m deep.

NESTLING

A baby bird that has not yet left the nest.

NEUROTOXIN

A toxin that impacts on the functioning of the nervous system. Some venomous animals, such as cone snails, sea snakes, stingrays, jellyfish and more, use this type of toxin to disable prey or predators.

NICHE
The function an organism plays within its environment.

NIDICOLOUS
Young birds that stay in their nest for a long time after they hatch, until they can fly. See also *altricial, nidifugous.*

The chicks of the **pied currawong** (*Strepera graculina,* right) are born blind and featherless and stay in the nest for an extended period. The nest is often targeted by the parasitic channel-billed cuckoo (*Scythrops novaehollandiae*), which lays its eggs there. When these eggs hatch, the mother currawong sometimes deserts the nestlings.

A pied currawong chick is NIDICOLOUS and reliant on parental care.

NIDIFUGOUS
Chicks that leave their nest immediately or soon after hatching. See also *precocial.*

Malleefowl (*Leipoa ocellata,* top left) are a threatened species of Australian bird. The male builds a large mound from leaf litter and soil, sometimes over 20 m wide. The female then lays eggs in a deep hole in the mound. Once the eggs hatch, the baby malleefowl dig their way out of the mound over several hours. They are immediately independent and leave the nest and the parents.

NOCTURNAL
Organisms that are most active at night. See also *crepuscular, diurnal.*

The **powerful owl** (*Ninox strenua,* right) is the largest species of owl in Australia. This nocturnal bird usually hunts for arboreal mammals, like the greater glider, but also occasionally eats bats, rats, rabbits, and other birds. The powerful owl is being affected by the widespread use of rat poison, which is making its way into the food chain.

NON-VASCULAR PLANT
Plants that do not have a vascular system with xylem and phloem but instead have more simple tissues to transport water. Moss and algae are both types of non-vascular plants.

Giant moss (*Dawsonia superba,* right) is a moss that can be found along the east coast of Australia in damp and wet forests. It is the tallest self-supporting moss in the world and can grow to about 60 cm tall.

Giant moss (below) is a NON-VASCULAR PLANT.

The powerful owl (bottom) is NOCTURNAL.

OCEANOGRAPHY
The study of anything relating to the ocean.

Mottlecah flowers (left) are revealed once the OPERCULUM falls away.

OCELLUS

The simple eye of some insects and other invertebrates, made up of light-sensitive pigments. Also used to describe the round, eye-like markings on some species, which mimic an eye to deter predators.

The **owl butterfly** (*Caligo sp.*, right) is known for the huge eyespots (plural ocelli) on its underside hindwings, which resemble the eyes of owls and are thought to distract would-be predators.

OLFACTION

The sense through which organisms detect smells.

OMNIVORE

Organisms that eat both plants and animals as a food source. See also *carnivore, herbivore, folivore, frugivore*.

Not only is the **whale shark** (*Rhincodon typus,* top right) the largest fish in the world, but it is also the largest omnivore. The whale shark survives on plankton, shrimp, small fish and seaweed.

OPERCULUM

A trapdoor-like structure in the shells of some snails and gastropods, both land- and water-based species. The operculum closes when the soft parts of the animal are retracted into the shell.

In botany, it is also used to describe the covering or 'cap' of a gumnut or other plant morphology that acts as a cap, hood or lid to protect the reproductive parts of the plant.

ORGANIC

Substances that are made from living material and are thus subject to decay over time.

ORGANISM

A living entity that can react to stimuli, reproduce and grow.

ORNITHOLOGY

The study of birds.

The whale shark (above) is the world's largest OMNIVORE.

Many butterfly species have OCELLI on their wings.

The eastern spinebill dines on ORNITHOPHILOUS plants and helps pollinate them.

Birds (above) that lay eggs (right) practice OVIPARY.

ORNITHOPHILOUS
Bird-loving or being pollinated by birds.

The **eastern spinebill** (*Acanthorhynchus tenuirostris,* top) relies on the nectar from ornithophilous plants as its main energy source. It is known to pollinate callistemon, grevillea and banksia species, among others.

OUTBREEDING DEPRESSION
A reduction in fitness that occurs when animals breed with others that are not closely related. See also *inbreeding depression*.

OVIPARY
When eggs are laid and embryos develop outside the mother's body, such as in snakes, lizards and birds. See also *ovovivipary, vivipary*.

OVIPOSITOR
A long, tube-shaped organ, often with the ability to pierce, used by some insects to lay eggs. Insects that have ovipositors include cicadas, grasshoppers and wasps.

OVOVIVIPARY
Embryos that develop within eggs and are kept inside the mother's body until they hatch. Animals usually use one reproductive method.

The **yellow-bellied three-toed skink** (*Saiphos equalis,* below) is found along Australia's east coast in New South Wales and Queensland. It is an unusual animal as it uses multiple methods of reproduction. A female, even in the same pregnancy, may use both ovipary and ovovivipary to produce young. One female was recorded laying three eggs and then giving birth to live young three weeks later. This skink is the first vertebrate species ever observed to do both in the same litter.

Yellow-bellied three-toed skinks can reproduce using OVIPARY and OVOVIVIPARY.

PALAEONTOLOGY

The study of prehistoric organisms through the examination of fossils.

Mistletoe is a PARASITIC plant that has a symbiotic relationship with the mistletoebird (right).

Siberia (above) was once part of PANGAEA.

The swift feather-legged fly (right) is a PARASITOID insect.

PANGAEA

A single supercontinent that existed 230 million years ago before it slowly broke apart into the continents that exist today. Pangaea formed from Euramerica, Gondwana and Siberia.

PARASITIC

When one organism survives by living in or on another organism. The parasite gains from the host, but the host doesn't gain anything from the parasite. See also *commensalistic, mutualistic, symbiotic*.

Mistletoe is a parasitic plant that grows on the trunks or branches of host plants. It relies on the host plant for water and food. Of course, just because they are parasites, doesn't mean they are bad! They're just occupying a niche within their environment.

Mistletoe also provides many benefits, including shelter for animals and nectar-rich, nutrient-dense flowers and leaves for birds and insects to eat. These plants have a symbiotic relationship with the **mistletoebird** (*Dicaeum hirundinaceum*, top). When their leaves fall to the ground, mistletoe plants add nutrients and a thick layer of leaf litter to the soil. There are more than 90 mistletoe species in Australia.

Some sharks (right) are capable of PARTHENOGENESIS.

Most wasps, like these yellowjackets, are PARASITOIDS or PARASITES of other insects.

PARASITOID

An organism that spends part of its life cycle as a parasite and part as a predator.

PARTHENOGENESIS

A form of asexual reproduction where an embryo develops from an egg that hasn't been fertilised by a male of the species.

PASSERIFORMES

An order of birds known as passerines, or perching birds, due to the anisodactyl arrangement of their toes. The order includes more than half of all bird species and all songbirds.

As well as being able to perch, songbirds have evolved special vocal organs to produce elaborate songs. While songbirds can be found across the entire world, they first evolved in Australia. **Magpies** (*Gymnorhina tibicen*, below right) are among Australia's most highly regarded songbirds and create complex vocal compositions. See also *anisodactyl*.

Scarlet honeyeaters (left) and magpies (below) are perching birds that belong to the order PASSERIFORMES.

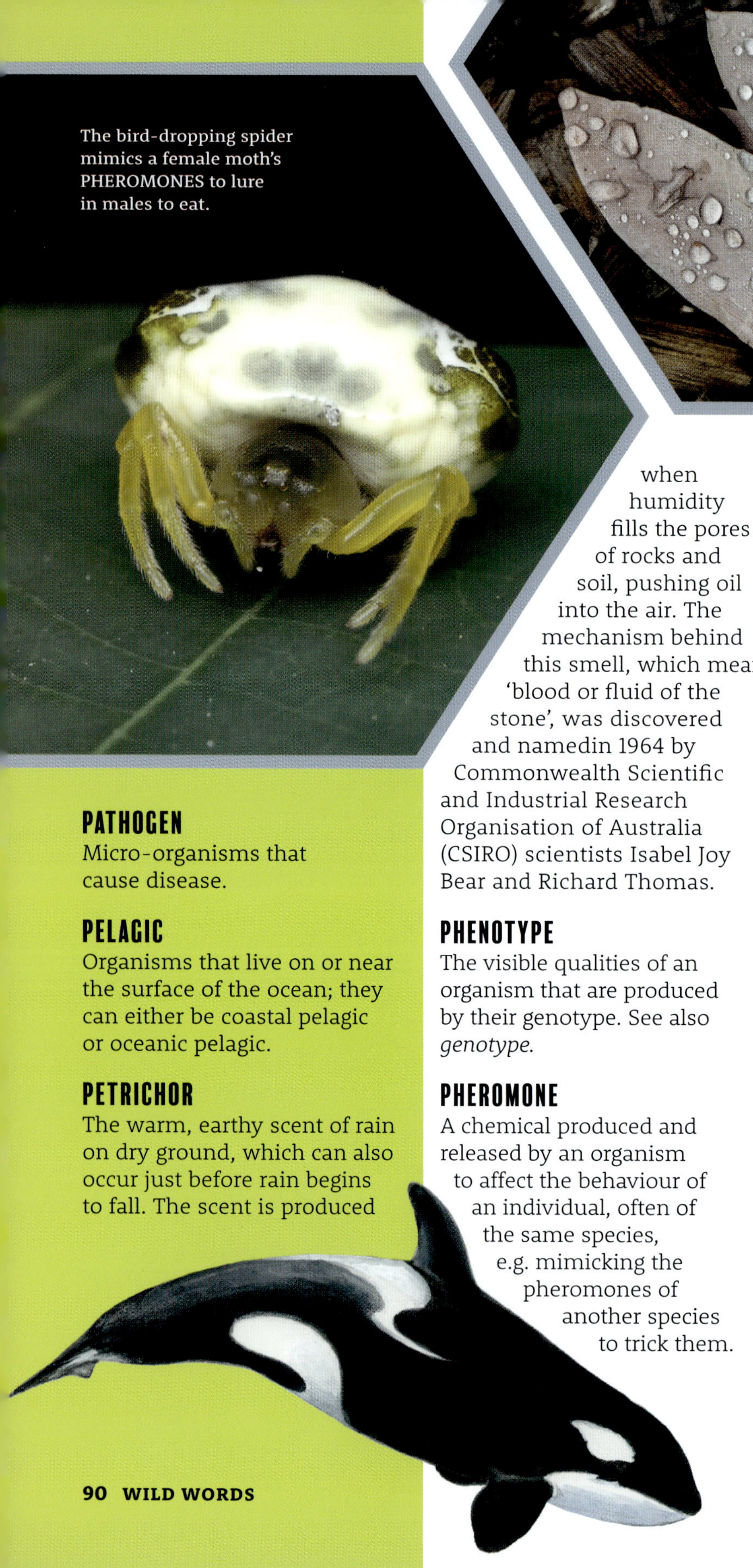

The bird-dropping spider mimics a female moth's PHEROMONES to lure in males to eat.

The scent of rain on dry ground is called PETRICHOR.

PATHOGEN
Micro-organisms that cause disease.

PELAGIC
Organisms that live on or near the surface of the ocean; they can either be coastal pelagic or oceanic pelagic.

PETRICHOR
The warm, earthy scent of rain on dry ground, which can also occur just before rain begins to fall. The scent is produced when humidity fills the pores of rocks and soil, pushing oil into the air. The mechanism behind this smell, which means 'blood or fluid of the stone', was discovered and namedin 1964 by Commonwealth Scientific and Industrial Research Organisation of Australia (CSIRO) scientists Isabel Joy Bear and Richard Thomas.

PHENOTYPE
The visible qualities of an organism that are produced by their genotype. See also *genotype*.

PHEROMONE
A chemical produced and released by an organism to affect the behaviour of an individual, often of the same species, e.g. mimicking the pheromones of another species to trick them.

The **bird-dropping spider** (*Celaenia excavata*, top left) from eastern and southern Australia releases a scent that mimics the pheromones released by female moths to attract mates. The male moths are lured in by the scent and eaten by the spider.

PHILOPATRY
The tendency for an organism to stay in or habitually return to a particular area. In some species, this can be more dominant in one sex. Male birds, for example, are more likely to be philopatric than females.

PHLOEM
Living tissue inside vascular plants that transports food throughout the plant. See also *xylem*.

PHORESY
A type of dispersal where one organism clings to the body of another before dropping off after a distance without being parasitic. For example, a beetle carrying mites.

An orca (left) is an oceanic PELAGIC sea mammal.

PHOTOPERIODISM

The physical response of organisms to changes in the length of nighttime. Many seasonal responses, including breeding, growing and migration, are determined by light and dark periods throughout the year.

PHOTOSYNTHESIS

The process used by plants and other organisms to create oxygen and energy from sunlight, water and carbon dioxide. See also *autotroph*, *chemosynthesis*, *heterotroph*.

PHYLETIC

Of or relating to the evolution of a species or group, particularly within a phylum.

PHYLUM

A ranking in Linnaean taxonomy that is below kingdom and above class.

PHYSICS

The study of matter, motion, energy and force and their behaviour across time and space.

PINNATE

Having leaves on the opposite sides of a stem.

PLANET

An object in space that orbits a star. It must be big enough that its gravity forces it into a spherical shape and pushes any other similar-sized objects away from its orbit.

Our solar system, the Milky Way, has eight planets: Mercury, Venus, Earth, Mars, Jupiter, Saturn, Uranus, and Neptune. Pluto is no longer considered a planet, but is one of five dwarf planets in the Kuiper Belt of our solar system.

Tongue orchids use mimicry to transfer their POLLEN.

PLANT

Plants are living organisms that generally produce their own food by using photosynthesis, have a waxy layer protecting them from drying out, and are sessile. Not all plants, however, use photosynthesis to produce their own food.

PLASTICITY

Capable of adapting in response to environmental stressors, changes or cues. Organisms with low levels of plasticity are more at risk of extinction.

The seeds of the Moreton Bay chestnut (below) are POISONOUS.

PLATE TECTONICS

A scientific theory describing giant plates that slowly move around underneath the land and the ocean. The movement of these plates creates mountains, volcanoes and earthquakes over time; it also split the larger supercontinent Pangaea into the continents that exist today. See also *Gondwana*, *Pangaea*.

POISONOUS

A toxic substance that can cause illness or death if eaten. See also *venomous*.

About 1000 Australian plants are poisonous to animals and humans. The black bean tree or **Moreton Bay chestnut** (*Castanospermum australe*, below left) is native to Queensland and New South Wales. It grows along the banks of rivers and in coastal rainforest. Between March and May, it produces large pods filled with poisonous seeds that can cause vomiting and diarrhoea if eaten raw.

POLLEN

A fine powder produced by plants for sexual reproduction. Pollen is carried from one plant to another by wind, birds, insects and other animals.

Tongue orchids (*Cryptostylis* spp., top left) have come up with a clever way of getting their pollen transferred from one flower to another. They release pheromones that mimic those given off by the female orchid dupe wasp (*Lissopimpla excelsa*). The male wasp is tricked into trying to mate with the flower, getting coated with pollen in the process. He then flies off to another flower that he thinks is a female and spreads the pollen. See also *pheromones*.

POLLINATOR

An animal that moves pollen from one flower to another. Pollinators are very important for the reproduction of flowering plants. Many animals are pollinators, including bees, flies, wasps, ants, bats, birds and possums. Bees, in particular, pollinate almost 90% of all the flowering plants on the planet. Australia has around 2000 species of native bee.

PRECIPITATION
Liquid or frozen water that forms in the atmosphere and falls to earth as rain, sleet or snow. Precipitation leads to annual snowfalls in the Australian Alps.

POLLUTION
The introduction of harm-causing materials into the environment. Pollution can ruin air, water and soil quality and is harmful to organisms, including to humans.

POLYGAMY
A reproductive strategy in which animals have more than one sexual partner at a given time. Kangaroos and other herd animals in which a single male mates with a 'harem' of females are polygamous. See also *monogamy*.

POPULATION
Organisms of the same species existing within a certain area.

POPULATION DYNAMICS
Studies that record and analyse the variables that cause equilibrium, increases or reductions in populations of organisms over time.

The blue-banded bee is a native POLLINATOR.

PRECOCIAL
Baby animals that can walk, run, hunt and survive independently almost immediately after being born. Herd animals that must quickly escape predators, such as horses and antelopes, tend to be highly precocious, as do reptile offspring, many of which disperse almost as soon as they hatch. See also *altricial, disperse* and *herd*.

Chickens (above) are considered very precocial birds. According to Macquarie University's Dr Birgit Szabo, individuals in each brood learn more flexibly than their parents.

Drosera sundews (right) are PREDATORY plants.

PREDATION

An interaction between organisms where one organism gets energy by feeding on another. Birds of prey such as the **osprey** (*Pandion haliaetus*, above) and sea eagles, swoop down to prey on smaller birds, mammals, reptiles and fish.

PREDATOR

In a predation interaction, the predator is the organism feeding on another. See also *prey*.

Plants can also be predators. In fact, Australia is home to the highest number of carnivorous plants on the planet, with more than 250 species. These carnivorous plants, such as **sundews** (*Drosera* sp., top right), lure in, trap, digest and eat invertebrates.

PREENING

A grooming behaviour that birds use to look after their feathers (above right). Preening both cleans the feathers and coats them in oil, making them resistant to water.

PREHENSILE

An appendage or organ, usually a tail, that is adapted for grasping and holding. The **brushtail possum** (*Trichosurus vulpecula*, right) uses its prehensile tail like a fifth limb to climb and jump between trees.

The horse-fly (left) uses a PROBOSCIS to feed on sap or blood.

PROKARYOTES called archaea turn the waters of some Australian lakes bright pink (below).

PREY
In a predation interaction, the prey is the organism being fed on. See also *predator*.

PROBOSCIS
A thin, flexible organ that forms part of the mouthpiece of some insects, including butterflies, flies and mosquitoes. It is mostly used for feeding.

PRODUCER
Organisms that are able to create their own food from inorganic substances. Producers provide food for consumers. See also *consumer*.

PROKARYOTE
Organisms without a complex cell structure that have cells without a nucleus bound in a membrane. This includes archaea and bacteria. See also *eukaryote*.

PYROPHILIC
In a literal sense 'fire loving'. It relates to plants that need fire to complete their life cycles or thrive after fire, e.g. some wattle species and carnivorous *Byblis* (right) species.

Q

QUADRUPEDAL
Animals that use four legs to move around.

The **bare-nosed wombat** (*Vombatus ursinus*) is a short-legged, muscular quadrupedal marsupial native to Australia.

QUASARS (below) are a part of a galactic nucleus.

The buff-breasted button-quail (right) was last seen in 1924.

QUAIL

Any of ten Australian species of small, short-tailed grassbirds in the order Galliformes, several of which are threatened species. Australia's species include button-quails, which are very distantly related to the quails of other continents. Among them are the critically endangered **buff-breasted button-quail** (*Turnix olivii,* top), a bird that has never been photographed.

QUARTZ

One of the hardest minerals on Earth, found in igneous, metamorphic and sedimentary rocks and used as a semi-precious stone or crystal.

QUASAR

Distant but extremely bright celestial objects that look like stars through a telescope but are actually a part of a galactic nucleus and are powered by massive black holes in space.

QUEEN

In highly social species – such as bees, ants and termites – the queen is the matriarch and the only fertile female that lays eggs.

QUENDA

A First Nations name for the western brown bandicoot (*Isoodon obesulus fusciventer*), a rabbit-sized marsupial found only in the South West region of Western Australia.

QUOKKA

Another Western Australian endemic, the **quokka** (*Setonix brachyurus*, below) is an unassuming and surprisingly friendly wallaby species that inhabits Rottnest Island, WA, where it is a major tourist attraction.

QUARTZ is a semi-precious stone that forms crystals.

RACEMES of the Western Australian Christmas tree.

R

RACEME
In botany, a raceme is a type of flower that is made up of a cluster of smaller, separate flowers spaced equidistantly by short stalks that adjoin a central stem.

RACHIS
In botany, the central stem of a plant. In anatomy, the central spinal chord of the body. And in ornithology, the central shaft of a feather.

RADIAL SYMMETRY
The body parts of a radially symmetrical animal are arranged symmetrically around a central axis. Starfish and sea urchins are examples of radially symmetrical animals. See also *bilateral symmetry.*

The Northern Pacific sea star (*Asterias amurensis*) is a radially symmetrical predator that has been accidentally introduced to Australian waters and threatens species including the critically endangered spotted handfish (*Brachionichthys hirsutus*).

RADICIVOROUS
An animal that is a root-eater.

RADIO TRACKING
Attaching a device that receives radio signals to an animal for the purpose of tracking its movements. This form of tracking involves using a directional antenna to detect the radio signal coming from the animal's transmitter, and following the signal to determine the animal's location. See also *GPS tracking.*

RADULA

A rasping tongue, used to scrape and cut food, which is found in most molluscs.

RANDOM SAMPLE

A subset of individuals or locations to study that each has an equal probability of being chosen. Random samples are used when there are too many organisms or the area is too large to study in its entirety.

RANIVOROUS

An animal that eats frogs. Some Australian snakes are primarily frog-eaters, including the common tree snake (*Dendrelaphis punctulata*).

RAPTORIAL

Predatory, or being a bird of prey (raptor) or like one.

REFUGE

An area that organisms hide in for protection. For example, an animal might find an inaccessible place to hide from a predator.

REPRODUCE

A biological process where organisms create new organisms by asexual or sexual reproduction.

REPTILE

Ectothermic, air-breathing vertebrates covered in a special skin of scales or bony plates (scutes). These animals include crocodiles, lizards, snakes and turtles. Australia has more than 860 species of reptile.

The sex of some reptiles' embryos, such as the central bearded dragon (*Pogona vitticeps*), can change depending on the ambient air temperature, which is known as temperature-dependent sex determination. If the eggs are incubated in an environment hotter than 32 °C, more eggs hatch as females. Increases in temperature therefore affect their survival.

Nankeen kestrels (left) are RAPTORIAL.

RADIO TRACKING can monitor a dingo's location (below).

Many species of snake (left) are RANIVOROUS.

Native ginger (pictured) grows from a RHIZOME.

The leatherback turtle is an R-SELECTED species.

Pelicans inhabit RIPARIAN zones.

RHIZOME

An underground plant stem that is usually horizontal and produces roots below, as well as shoots above, that create a new plant. **Native ginger** (*Alpinia caerulea*, above), like other gingers, grows from a rhizome.

RIPARIAN

The land and plant communities adjacent to rivers, streams and wetlands. These areas help to reduce erosion and filter pollutants before they reach the water. They also provide important habitat and corridors for wildlife to move through.

R-SELECTION

R-selected species produce a high number of offspring that receive little or no parental care and have a low survival rate.

The **leatherback sea turtle** (*Dermochelys coriacea,* left) uses an r-selection strategy: producing numerous eggs and providing no parental care once they hatch. Female leatherbacks return to the same tropical beaches in northern Australia where they hatched to lay their own eggs. The female digs a nest in the sand and lays about 110 eggs before returning to the ocean. She does this up to nine times each breeding season.

These turtles are threatened with extinction due to over-development of beaches, water warming and pollution, competition with invasive species, and netting and overfishing. Light pollution near beaches can also affect turtle hatchlings, as they use moonlight to help orient themselves on hatching. See also *K-selection*.

SCENT MARKING

Smelly secretions – such as faeces, urine or saliva – that an animal deliberately leaves to communicate a message to other animals.

Spotted-tailed quolls (*Dasyurus maculatus*, pictured) poo on flat rocks or in high places known as 'latrines'. Latrines function like public poop 'message boards' for quolls, letting others know who is in the area, what they have eaten, their sex, and even whether they're ready to mate. Bare-nosed wombats (*Vombatus ursinus*) also desposit their cube-shaped poop atop logs, sticks and rocks at nose height to ensure other wombats smell them.

The yellow-footed antechinus (right) is SEMELPAROUS.

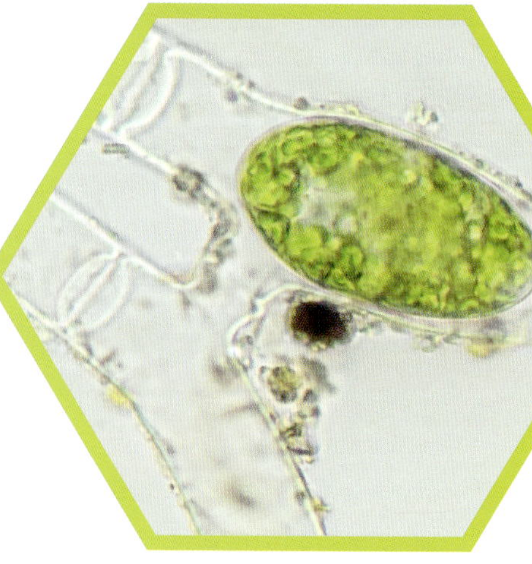

The handsome bush-pea (top) is SCLEROPHYLLOUS.

SEXUAL REPRODUCTION in *Spirogyra* sp. (above).

SCIENCE

The pursuit and application of knowledge by using observation, experimentation, and testable theories.

SCINTILLA

A very tiny, or minute, amount or trace of something.

SCINTILLATE

To twinkle, sparkle or emit light in flashes. The scintillation we see when we look at stars is due to starlight refracting (bending) as it passes through our atmosphere.

SCLEROPHYLLOUS

Having hard, short and often spiky leaves that are well-adapted to preventing water loss. Australia has many sclerophyll forests that are made up mostly of eucalypts and other drought-tolerant plant species.

SEMELPAROUS

A plant or animal that only breeds or reproduces for a single mating season in a lifetime. See also *iteroparous.*

The males of some small dasyurid species, including the **yellow-footed antechinus** (*Antechinus flavipes,* top) and the red-tailed phascogale (*Phascogale calura*), expire after their first mating season. The males of these marsupials spend so much time mating in their first breeding season that they become malnourished, stressed and die.

SENESCENCE

The process by which most organisms deteriorate with age.

SESSILE

Attached to a substrate and not able to move independently. See also *motile.*

The **waratah anemone** (*Actinia tenebrosa*, below) is a brownish-red sea anemone found throughout the coastal waters of southern Australia. These anemones are sessile. They attach to rocks and use their tentacles to snare food that floats past.

SEXUAL REPRODUCTION

A type of reproduction that involves the fusion of cells from different individuals to create offspring. See also *asexual reproduction.*

SEXUAL SELECTION

A genetic selection process driven by mate choice rather than direct survival. The evolution of particular traits occurs due to the mate preference of a particular sex. See also *artificial selection*, *natural selection*. **Peacock jumping spiders** (*Maratus* sp., top right) have brightly coloured abdomens that they shake in a courtship display to attract females. As females choose the males that have the most impressive colours, over time, the diversity and vibrance of male colouration across the genus has increased.

The peacock jumping spider (above) uses SEXUAL SELECTION when choosing mates.

A waratah anemone is SESSILE.

SIBLICIDE

When an animal kills its siblings. The **laughing kookaburra** (*Dacelo novaeguineae*, right) is an iconic Australian bird endemic to eastern Australia. It has been introduced to other areas, including Tasmania and Western Australia. The female lays three eggs, but the two chicks that hatch first usually peck the third to death to increase their access to food.

The laughing kookaburra (below) commits SIBLICIDE.

SOLAR ECLIPSE
When the Moon moves between the Sun and the Earth, temporarily obscuring the Sun.

A solar eclipse only happens at the new moon, and it can only be seen during the day. If it is a partial eclipse, only a part of the Sun will be covered, but if it is a total eclipse, darkness descends temporarily. See also *lunar eclipse*.

A SOLAR ECLIPSE can only be seen when there is a new moon.

SOLAR SYSTEM
Everything in space that is drawn by gravity around our Sun's orbit. This includes planets, dwarf planets, moons, asteroids and nebulae (pictured).

SONG
A pattern of sound that is created by specialised organs. Animals use song as a way of communicating with each other.

The **Australian magpie** (*Gymnorhina tibicen*, right) has a complex range of calls, some of which can vary in pitch by four octaves. The magpie's familiar melodious warble can be heard in many backyards, but it is also a talented mimic that can reproduce the sounds of dogs, cats, humans, other birds, as well as of ambulances and emergency vehicles.

SPECIES
A group of related organisms that share common characteristics and are reproductively distinct from other organisms.

About 1.2 million species on Earth have been described, but scientists estimate there may be some 8.7 million species in total.

SPELEOLOGY
The scientific study of, or exploration of, caves.

SPINESCENT
Having or bearing spines, or becoming spine-like.

SPORE

A reproductive cell that can turn into a new individual organism without fusing with another cell. This method of asexual reproduction is used by algae, bacteria, fungi and plants.

The **puffball fungus** (*Lycoperdon perlatum*, right) releases its spores in a cloud that looks like a puff of smoke; usually, this release is triggered by the impact of a raindrop or a passing animal.

STAR

Bodies of gas that produce energy and light that can be seen from Earth as a bright point in the sky. The Sun is the closest star to Earth.

STERILE

An organism that is unable to reproduce, land where plants cannot grow, or an environment that is devoid of any living organisms.

STRESS

A physiological condition brought on by excessive environmental or psychological pressures. Stress can be short-term or prolonged.

STYGOFAUNA

Organisms that live under the ground in groundwater, fissures, vugs or caves.

SUBSTRATE

The surface or material that an organism lives or grows on.

SUSTAINABLE

Able to be maintained at a certain rate or level. Sustainability is often used to describe the goal of humans existing on Earth for a continued period of time without causing major environmental problems such as extinctions, pollution or resource scarceness.

SYMBIOTIC

A close relationship between two species that benefits one or both, including mutualistic and parasitic relationships. See also *commensalistic, mutualistic, parasitic.*

SYMPATRIC

Species that live in similar habitats or whose habitats overlap. See also *allopatric.*

SYMPATRIC SPECIATION

When organisms evolve genetic differences over time although they live in the same area and aren't separated by geographic barriers. See also *allopatric speciation.*

A TARDIGRADE (top) is an extremely tough organism.

The palm cockatoo (above) is known to be TERRITORIAL.

T

TAPETUM LUCIDUM

Reflective tissue behind the retina of some animals' eyes that aids night vision and creates eyeshine.

TAPHONOMY

The study of how organisms decay and become preserved as fossils or other materials, in the paleontological record.

TARDIGRADE

Microscopic eight-legged organisms also known as water bears or moss piglets, the largest of which grows to just 0.5 mm. Earth has about 1300 species of **tardigrade** (top left). They are often found living in the films of water surrounding moss, but they can be found anywhere from the tops of mountains to the deep sea. Tardigrades are extremophiles that can survive temperatures as low as –272 °C and as high as 151 °C. They can be starved of nutrients, oxygen and water and still survive!

TAXONOMIC RANK

A way of categorising species. There are eight major ranks: domain, kingdom, phylum, class, order, family, genus, and species.

Below is the taxonomic rank of a short-beaked echidna (*Tachyglossus aculeatus*).

Domain: Eukaryote
Kingdom: Animalia
Phylum: Chordata
Class: Mammalia
Order: Monotremata
Family: Tachyglossidae
Genus: *Tachyglossus*
Species: *aculeatus*

TAXONOMY

The study of naming, defining and classifying organisms.

TERMITOPHILE

Species that spend part of their lives closely associated with termites; they are often insects living in termite mounds.

TERRITORIAL

A behaviour animals use to defend their territory. Species often use territorial behaviour during their breeding season.

The **palm cockatoo** (*Probosciger aterrimus*, left) lives on the northern tip of Queensland. It drums sticks against its hollow tree nest to alert neighbouring males to its claim on a territory.

More than 1700 species and ecological communities are at risk of extinction in Australia. The major threats to these species and communities include habitat destruction, invasive species and climate change. Today, fewer than 2000 critically endangered **mountain pygmy-possums** (*Burramys parvus*, above) remain in the wild. Their main food source. the bogong moth (*Agrotis infusa*, right), is also on the IUCN Red List of Threatened Species, making the survival of this pygmy-possum even less certain.

TERRITORY

An area defended by an organism, or group of organisms, that they use for mating, nesting, sleeping or feeding.

THREATENED SPECIES

Species or communities that are at risk of becoming extinct in the near future. Several categories are used to classify threatened species, depending on their risk. Threatened species include those that are vulnerable, endangered and critically endangered.

The mountain pygmy-possum (above) is a THREATENED SPECIES.

The bogong moth has a TAXONOMIC RANK.

TRANSLOCATION has helped conserve the greater bilby (below).

TIDE
The rise and fall of sea levels caused by gravitational forces exerted by the Moon and the Sun. Tides change roughly every 12 hours.

TOPOGRAPHY
The physical features of an area of land, such as mountains, hills, rivers, lakes and valleys.

TORPOR
A method of saving energy, whereby animals spend periods of time with very low body temperatures and levels of activity. While similar to hibernation, torpor can happen daily or at any time of year. See also *aestivation*, *hibernation*.

The eastern pygmy-possum (*Cercartetus nanus*) inhabits the east coast of Australia, from Queensland to South Australia and throughout Tasmania. It holds the title for the longest periods of torpor of any mammal in the world and is capable of spending a whole year in a less-active state.

TOXIN
An organic poison produced by an organism. Toxins are produced by both poisonous and venomous organisms. Toxins are generally used by predators, either to catch food or for defence against being attacked or eaten.

TRANSECT
A path of a set distance that a researcher travels along to count and record the objects or organisms along it. Transects are used to determine the number of organisms in an area.

TRANSLOCATION
The capture, transportation and release of a species to another area. Translocations are sometimes used in Australia to conserve threatened species and can help increase the genetic diversity of an at-risk population. The **greater bilby** (*Macrotis lagotis*, left) was translocated back to New South Wales after being absent for almost 100 years. The reintroduced bilbies are protected in a large fenced area that is free of introduced feral predators.

TROPHIC CASCADE
A trophic cascade occurs when an apex predator is either added to or removed from an area, which leads to changes in the number or behaviour of prey species. In turn, this alters the density or behaviour of other species, a seismic shift that 'cascades' through a food chain and affects the ecosystem.

TURBIDITY
A measure of the cloudiness of a fluid caused by suspended particles within it. A turbidity measurement is often used when assessing water quality.

TWITCHER
A birdwatcher that goes to great lengths to find species they haven't seen before. Twitchers keep lists of all the bird species they have seen.

The broad-headed snake (right) produces a TOXIN that it injects as venom.

U

The koala is an UMBRELLA SPECIES.

ULGINOUS

Growing in swamps or preferring very wet ground. The floating water primrose (*Ludwigia peploides*), although native to Australia, is an ulginous aquatic plant that can easily become weedy and take over wetlands and swamplands.

ULTRADIAN

In chronobiology, this is a physiological cycle that takes place for longer than an hour and reoccurs within the space of a day.

UMBEL

In botany, an umbel is when a cluster of flowers springs from a single centre. In Australia, most umbelliferous plants are from the Apiaceae (celery) family, including the blue devil (*Eryngium ovinum*), which is endangered in South Australia.

UMBRELLA SPECIES

A species that is chosen when making conservation decisions and actions because protecting that species will indirectly help many other species.

The **koala** (*Phascolarctos cinereus*, above) is used as an umbrella species in Australia. By protecting koala habitat, all other plants and animals that use the same habitat will also be protected.

UNICELLULAR ORGANISM

An organism that is made up of only a single cell. Unicellular organisms can be either prokaryotic, such as bacteria and archaea, or eukaryotic, such as some algae and fungi. See also *multicellular organism*.

Algae is a UNICELLULAR ORGANISM.

UNIVERSE

Space, time and everything it includes, plus all matter, energy, galaxies, black holes, stars and planets, including the Earth.

V

A VECTOR can infect burrowing bettongs.

VASCULAR PLANT

Vascular plants are a large group of plants that have specialised vascular tissues, xylem and phloem to transport water and food. This group of plants includes angiosperms, gymnosperms and ferns.

The **mountain ash** (*Eucalyptus regnans,* above) is a species of vascular plant found in south-eastern Australia. These giant trees are the tallest flowering plants on Earth. The tallest living mountain ash, named Centurion, is in Tasmania and is more than 100 m tall.

VECTOR

An organism that carries a disease-causing organism from an infected individual to a healthy individual. Toxoplasma (*Toxoplasma gondii*) is a parasite that can infect most mammals, including humans. Cats are a vector for this parasite, but it must enter their gut to reproduce sexually. Cats then spread the parasite via their faeces. Toxoplasma can cause behavioural changes and death in infected mammals, such as the **burrowing bettong** (*Bettongia lesueur*, top right) and eastern barred bandicoot (*Perameles gunnii*). It can also cause illness and behavioural changes in humans.

VELLUS HAIR

A type of very fine, almost downy, soft and pale non-pigmented hair that covers the skin of mammals, including humans, and helps regulate body temperature by thermal insulation.

VENATION

The pattern or arrangement of veins in a leaf or on an insect's wing – like the wings of the **Sydney hawk dragonfly** (*Austrocordulia leonardi*, below).

The common lionfish is VENOMOUS.

VENOMOUS

An animal that has a venom delivery system capable of injecting a toxin into another animal by biting, stabbing or stinging. See also *poisonous*.

Australia is home to the world's most venomous snake, the inland taipan *(Oxyuranus microlepidotus,* right). Although no recorded human deaths have been attributed to this species, a single bite from this snake contains enough venom to kill up to 250,000 mice and has the potential to kill a human within 45 minutes. Fortunately, the inland taipan, as its name suggests, lives in the arid interior of the country, so humans very rarely encounter this species.

VENTRAL

Of, pertaining to, or situated on the underside of the body or the belly or abdomen.

VERTEBRATES

Animals that have a backbone or spinal column. Fish, amphibians, reptiles, birds and mammals are all vertebrates. See also *invertebrates*.

The inland taipan (above) is considered the world's most VENOMOUS snake.

VERTEBRATES, like this crocodile (below), all have a spinal column, also known as a backbone.

VESTIGIAL

A structure or organ that has lost its original function and is reduced in size or ability. For instance, the vestigial hindlimb buds that some python species still possess, the pelvic girdle of snakes, or the wings of penguins (below) or ratites such as the **emu** (*Dromaius novaehollandiae*, below right). In humans, the appendix is thought to be vestigial.

VICARIANCE

When a sizable group of organisms from a population are separated due to a geographic barrier, usually caused by a geological event such as the formation of an abyss, an earthquake or a new river.

VIVIPARY

When embryos develop inside the female, which then gives birth to live young. Bougainville's skink (*Lerista bougainvillii*) is thought to be mostly oviviparous on the mainland, but two distinct island populations have been found to be viviparous.

VOCALISATION

Sounds produced by an animal from air passing through the vocal cords. Animals 'vocalise' to communicate messages to other animals.

The blue whale (*Balaenoptera musculus*, pictured) is not only the largest animal in existence today but also the largest animal to ever have lived! It also produces the loudest vocalisations – loud whistling calls to others of its species, which can be up to 188 dB. That's louder than a jet engine or a grenade exploding!

VOLCANO

An opening in the crust of the Earth that allows lava, rocks and steam to erupt from below the surface. Volcanoes are usually found near the edges of tectonic plates.

Active volcanoes are rare in Australia because our continent does not sit on top of any tectonic plate boundaries. Only two active volcanoes are now found in Australian territory; both are about 4100 km south-west of Perth, on Heard Island (left) and the nearby McDonald islands. However, Australia does bear evidence of earlier fiery explosions, notably at Undara Volcanic National Park in Queensland, where the world's longest lava tubes remain, and **Blue Lake** (below) in South Australia, which fills the crater left by an earlier volcano at Mount Gambier.

VOLCANOLOGY

The study of volcanoes, lava, magma and related geological activity.

VORTEX

Derived from the Latin word *vortere* ('to turn'), a vortex is a number of particles moving in a rapid circular motion around a central axis. Species that have too little genetic diversity left to avoid inbreeding are said to be in an 'extinction vortex'.

WADER

Any species of bird, usually long-legged, that wades in waterways searching for food. Australian waterways are havens for many migratory and permanent wading species, among them the critically endangered **eastern curlew** (*Numenius madagascariensis,* left), the **black-necked stork** (*Ephippiorhynchus asiaticus,* top), which is also known as the jabiru, and the brolga (*Grus rubicunda*).

WADI

The bed of a watercourse in an arid area, often in a valley, which usually remains dry outside of the rainy season.

WANING

A cycle of the Moon during which it has only a small, and diminishing, part of its surface lit up and visible, so that it seems reduced in size.

WATER TABLE

An underground boundary between surface soil and groundwater. Groundwater is water that has penetrated the surface of the Earth and fills spaces between sediments and rocks.

Australia's Great Artesian Basin is the world's largest

subterranean freshwater resource. More than 400 fascinating Australian stygofauna species live in groundwater; most are crustaceans, but some are eels, fish, snails and worms. See *stygofauna*.

WAVELENGTH

A measurement of the distance between two points – mostly from crest to crest – in a single cycle of a wave. Wavelength is represented symbolically by the lower case Greek letter lambda (λ).

WAXING

When the Moon appears to be increasing in size towards a full moon as more of it is illuminated and visible.

WEATHER

The state of the atmosphere, which includes the temperature, level of humidity, precipitation, wind and cloud cover. Weather varies day to day.

Blackberries were introducted to Australia and are now a WEED.

Australian sea lions (top) and wallabies (above) are native WILDLIFE in Australia.

WEED

In a natural setting, weeds are usually plants that have been introduced to a region or country, have gone wild, and are now causing environmental issues.

Blackberry (*Rubus fruticosus*, above) has invaded many parts of Australia. It grows rapidly and crowds out native plants, stopping new native seedlings from thriving. Blackberry was deliberately introduced by European settlers in the mid-1800s for its delicious fruit, but it quickly became a serious agricultural issue and began to outcompete native floral species. Australia currently has at least 32 weeds of national significance (WONS), meaning all Australians are obliged to remove or report them, for biosecurity reasons.

WILDLIFE

Wild, undomesticated native animals. Kangaroos, wallabies, wombats, echidnas, and **Australian sea lions** (*Neophoca cinere*a, top left) are examples of Australian mammalian wildlife. Camels, goats and pigs are not considered wildlife, as they are introduced or escaped (feral) domestic animals.

WILD TYPE

The most common or typical form of a plant or animal that exists in the wild. In the field of genetics, this is the purest, least mutated strain of a species.

X
XANTHOCARPOUS
Bearing yellow fruit.

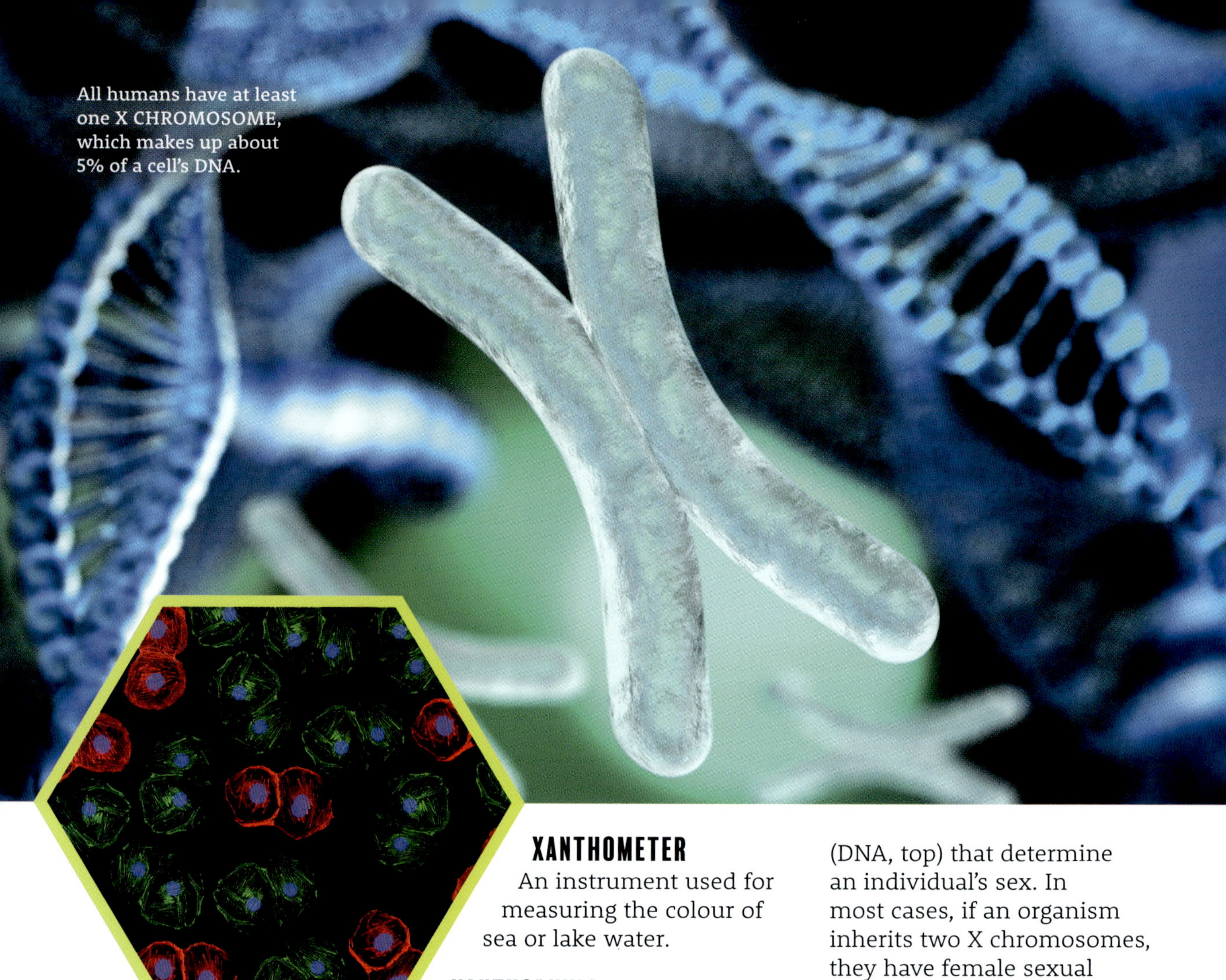

All humans have at least one X CHROMOSOME, which makes up about 5% of a cell's DNA.

Gene-edited DNA from pigs (above) is XENOGENEIC to humans.

XANTHOMETER

An instrument used for measuring the colour of sea or lake water.

XANTHOPHYLL

A class of carotenoid pigments that result in yellow, orange and red colour of flowers, fruits, vegetables and leaves, as well as the colours of some animals.

XANTHOSPERMOUS

Possessing yellow seeds.

X CHROMOSOME

In most mammals (and some other species), the X chromosome is one of two sex chromosomes made of deoxyribonucleic acid (DNA, top) that determine an individual's sex. In most cases, if an organism inherits two X chromosomes, they have female sexual characteristics. See also *Y chromosome*.

XENOBIOTIC

A substance or item that is foreign to the body or to an ecological system.

XENOGAMY

In botany, the fertilisation of one flower or plant by pollen from another, or cross-fertilisation; for instance, rockmelons (left) and honeydew melons are both members of the cucurbit family and will cross-pollinate if grown near each other. In biology, the term means the fertilisation of an organism by the fusion of an egg from one individual with a sperm of a different individual.

XENOGENEIC
Relating to or involving tissues or cells from a different species.

A xenogeneic transplant is a procedure involving the transplantation, implantation or infusion into a human recipient of live cells, tissues or organs from a non-human animal source, e.g. pig cells to help restore lost tissue function and heal wounds in a human, or a modified pig organ transplanted into a human.

XENOGENESIS
The supposed generation of offspring entirely unlike the parent.

XERANSIS
Drying up; the gradual loss of tissue moisture.

XERARCH
Growing in dry places, or a plant community that originates in dry surroundings, such as desert or sand dunes.

XERIC
Dry and lacking in moisture or relating to an environment characterised by aridity.

XEROPHILOUS
Growing in or adapted to hot and dry regions. Australia's many everlasting daisy species in the Asteraceae family are xerophilous, as are the mulla mulla (*Ptilotus* spp., above) that spring to life in the deserts.

Beavers are XYLOPHAGOUS mammals of the Northern Hemisphere.

XEROPHOBOUS
Unable to survive drought.

XEROPHYTIC
A plant adapted to withstanding drought.

Many Australian plants have adapted to the harsh, dry climate. Succulents such as pigface (*Carpobrotus rossii*) have evolved swollen water storage organs so that when it rains, they can quickly absorb water and keep it in reserve for drier times.

XIPHOPHYLLOUS
Having sword-shaped leaves, such as in the **giant spear lily** (*Doryanthes palmeri,* below left).

XYLEM
The living tissue inside vascular plants that transports water throughout a plant. See also *phloem*.

XYLOGENOUS
Growing on wood. For example, **oyster mushrooms** *(Pleurotus ostreatus, right)* grow on tree trunks.

The giant spear lily (left) is XIPHOPHYLLOUS.

XYLOPHAGOUS
Herbivorous animals that primarily eat wood, including beavers, wood beetles, the Australian wood cockroach and the larvae of some insects, such as the witchetty grub (*Endoxyla leucomochla*), which bores into the roots of the wanderrie wattle (*Acacia kempeana*).

XYLOTOMOUS
Cutting into or boring into wood, as with certain insects.

Y

Y CHROMOSOME

In most mammals and some other species, the Y chromosome is one of two sex chromosomes that determine an individual's biological sex. An individual that inherits an X and a Y chromosome from its parents will generally have male characteristics. See also X *chromosome*.

In many species, biological males and females look similar, but in others, sexual dimorphism occurs. In primates, for which X and Y chromosomes determine biological sex, some of the most obviously sexually dimorphic examples are seen in lowland gorillas (*Gorilla gorilla*), in which males are 2.3 times the weight of females, and in **mandrill monkeys** (*Mandrillus sphinx*, pictured), in which the male (right) is obviously larger than the female (below) and has much more vivid facial markings.

YARAK

A hyper-alert state where a bird is in good flying or hunting condition, such as a hawk or eagle (left).

YARDANG

A sharp rock ridge up to 6 m high, formed in a desert by wind erosion parallel to prevailing winds (right).

YATE

Eucalyptus cornute – a species of eucalyptus tree with strong wood that is native to south-western Western Australia.

YIKKER

The sharp little cries of a bird or animal, or to squeal or squeak sharply and repeatedly, as for baby birds in a nest (above).

YOUSTER

To fester.

Z

ZALAMBDODONT

Having molar teeth with V-shaped ridges. Australia's two marsupial mole species in the Notoryctidae family, for example, are zalambdodontic.

The number of hours of sunlight in a day form a ZEITGEBER.

ZEITGEBER

A rhythmically occurring event, such as light and its impact on the circadian clock, that cues an organism's biological rhythms.

ZENOGRAPHY

Study of the planet Jupiter.

ZINCIFEROUS

Yielding or containing zinc.

ZOOARCHAEOLOGY

Also known as faunal analysis, this branch of archaeology studies the remains of animals – such as teeth, bones and shells – that are found at archaeological sites.

ZOOCHORY

The spread of plant seeds or spores by animals, including birds, beetles and furred animals. In Two Peoples Bay Nature Reserve, Western Australia, biologists observed three mammal species eating the **Australian bluebell** (*Billardiera fusiformis,* bottom left), and thus dispersing its seeds. This type of zoochory is known as 'endozoochory'. The bush rat (*Rattus fuscipes*), the critically endangered Gilbert's potoroo (*Potorous gilbertii*) and the quokka (*Setonix brachyurus*) all dispersed the seeds via their digestive tracts. A 2005 study found their actions increased germination by 58%!

ZOOGAMY
Sexual reproduction of animals.

ZOOGONOUS
In zoology, giving birth to live offspring, rather than laying eggs; in botany, seeds or fruit that sprout before they fall from the parent plant.

ZOOIDS
Single animals that make up part of a colonial animal. Coral and jellyfish are both examples of zooids.

ZOOPLANKTON are often minuscule or microscopic.

The bluebottle (*Physalia utriculus*) is a specialised colony of different organisms. It can be found washed up on beaches across Australia. Specialised polyps make up the float, tentacles, digestive system and reproductive system. Its tentacles can also inject a powerful venom into its prey – and into any human unlucky enough to come into contact with it.

ZOOLOGY
The study of the behaviour, structure, classification and distribution of animals, both living and extinct, and how they interact within their ecosystems.

ZOONOSIS
An infectious disease that can be transmitted from an animal to a human host through contact (e.g. rabies or Hendra virus) or from humans to animals. Giardiasis is an infection caused by the parasite *Giardia lamblia*. Humans can catch it from coming into contact with and ingesting infected animal faeces. It can cause severe diarrhoea, nausea and fatigue. One of the most common ways people get infected is by drinking contaminated water, which is why it is always a good idea to boil any water collected from a river or creek before you drink it.

ZOOPLANKTON
The animal or animal-like constituent of plankton, comprised mainly of freely floating protozoa, small crustaceans (e.g. krill or copepods), fish eggs and larvae.

ZOOTOXIN
Any poison derived from an animal, including from bees and wasps, cane toad poison, lamprey slime, stingray or snake venom –as in the **eastern small-eyed snake** (below).

IMAGES

FRONT COVER & TITLE PAGE: Holger Grybsch/PB; Alexander Lesnitsky/PB; James McKinnon/AG; Rose Colbeck/AG.

BACK COVER: Alex Cooper Photography/SS; Scott plume/SS; Anna Lopatina/D; Rose Colbeck/AG; Marje Crosby-Fairall/AG; Michael Burleigh/AG.

AdobeStock = AS; Australian Geographic = AG; D = Dreamstime; Pixabay = PB; Shutterstock = SS; Unsplash = US.

p.1: Bruce Thomson/auswildlife.com. **SECTION A: p.3** Holger Grybsch/PB; AlexanderLesnitsky/PB. **p.4** David Clode/US. **p.5** Butterfly Hunter/SS. **p.6** Diego Grandi/SS. **p.7** Rostislav Stefanek/SS; Isselee/D; Maple Ferryman/SS. **p.8** Alex Gimenez/SS; john austin/SS. **p.9** Mike McCoy/AG; Nick Rains/AG; belizar/SS. **p.10** Jason Edwards/AG. **p.11** Tamara Kulikova/SS; Beth Ruggiero-York/SS; Joseph Pérez/US; Mitch Reardon/AG. **p.12** Craig Lambert Photography/SS; © Bjoern Wylezich/SS. **p.13** Christel SAGNIEZ/PB; Martchan/SS;Kevin Stead/AG. **p.14** © Holger Grybsch/PB; Ego Guiotto/AG; ISKYDANCER/SS. **p.15** Danijela Maksimovic/SS; Rob D the Pastry Chef/SS. **p.16** Sahara Frost/SS; Ken Griffiths/SS. **p.17** Bruce Thomson/auswildlife.com; Kevin Stead/AG. **p.18** Potapovpaladin/SS; SKT Studio/SS. **p.19** Pexels/PB; © Christine Jones/@Mycohuman. **p.20** Eric Isselee/SS; Robert Eastman/SS; Marty R Hall/SS;Kezza/SS. **SECTION B: p.21** Irma/AS; **p.22** Ethan Daniels/SS; Robert Buchal/SS; Ольга Исхакова/AS. **p.23** John A Anderson/SS; Norrawith/SS; Khaiboy/SS. **p.24** Taras Vyshnya/SS; Susan Flashman/SS; **p. 25** Konan Farrelly-Horsfall/iNat; Esher Beaton/AG. **p.26** manuel/AS; Iryna Mylinska/SS; Juice Flair/SS. **p.27** Ken Griffiths/AS. **p.28** ShutterSparrow/SS; Phonlamai Photo/SS; jaroslava V/SS. **p.29.** Artfully Photographer/SS; Lindsey Lu/SS; James McKinnon/SS. **p.30** PiotrKrzelak/SS.

SECTION C: **p.31:** Rattiya Thongdumhyu/SS. **p.32:**Kenny CMK/SS; Dominyk Lever/SS; Ton Bangkeaw/SS. **p.33**: Justin Gilligan/AG; AGCreations/SS; Ozja/SS. **p.34:** Lukas Tennie/US; Divelvanov/SS; Leena Robinson/SS. **p.35:** Anavarroc/D; Geoff Childs/SS. **p.36:** Justin Gilligan/AG; **p.37:** ansiemartin/SS; MPIX/SS; Chris Andrews Fern Bay/SS. **p.38:** Dirk Kotze/SS; Kevin Stead/AG; Alex Coan/SS; Louise Saunders/AG. **SECTION D**: **p.39:** Mike Langford/AG. **p.40:** Steve Wilson/AG; Anne Powell/SS; Eric Isselee/SS. **p.41:** Chatchai Somwat/SS; Joey Kyber/US; Sunet Suesakunkhrit/SS. **p.42** Lukas_Vejrik/SS; ice_blue/SS. **p.43:** Jemma Cripps/SS; Chones/SS; Luke Shelley/SS. **SECTION E: p.44** Rod Scott/AG; OlgaLi/SS; scott plume/SS. **p.45** Johan Larson/SS; Bradley Blackburn/SS. **p.46** Kevin Stead/AG. **p.47** BRONWYN GUDGEON/SS; ausnative/SS; Vera Larina/SS; GoodFocused/SS. **p.48:** Mari May/SS; Quentin Chester/AG. **p.49** Raphael Bick/US; Rod Scott/AG. **SECTION F: p.50** Brayden Stanford Photo/SS; Tee Wong/SS; slowmotiongli/SS. **p.51:** Redchanka/SS. **p.52:** Kevin Stead/AG; Imogen Warren/SS; Anna Lopatina/D. **SECTION G: p.53:** 3 Travelers/SS. **p.54**: Bildagentur Zoonar GmbH/SS; Ego Guiotto/AG; mic_136/SS. **p.55** Thomas Lusth/SS; Pascal van de Vendel/US; Minakryn Ruslan/SS. **p.56:** David Bristow/AG; Debra Anderson/SS. **SECTION H: p.57** 2j architecture/SS. **p.58:** pR13S7/SS; Michael Burleigh/AG. **p.59:** Benny Marty/SS; Ken Griffiths/SS; Jene Smu/SS. **p.60:** bogdan ionescu/SS; Wolfilse/SS; Ego Guitto/AG. **SECTION I: p.61:** Aleksandr Rybalko/SS. **p.62:** Santanu Banik/SS; Samantha Haebich/SS; Helen J Davies/SS. **p.63:** I Putu Krisna Wiranata/SS; Olivia Parsonage/AG; slowmotiongli/D. **p.64:** Nick Rains/AG; Slowmotiongli/SS; AlenThien/SS. **SECTION J: p.65:** sirtravelalot/SS. **p.66:** Chris/AS; Kristian Bell/SS; Ermek Oksana/SS. **p.67:** Boris PamikovSS; Nick Rains/AG. **SECTION K**: **p.68:** Lewis Burnett/AG; phototrip.cz/SS. **p.69:** Ondrej Prosicky/SS; Nicolas Day/AG; K.A. Willis/SS. **SECTION L**: **p.70:** ImageBank4u/SS;Chrissie Goldrick/AG; KOOKLE/SS. **p.71:** GagliardiPhotography/SS; I Wayan Sumatika/SS; Lewis Burnett/AG. **p.72:** David Dare Parker/AG; Wirestock Creators/SS; AstroStar/SS. **SECTION M: p.73:** Ralf Lehmann/SS. **p.74:** Rapeepong Phataiwunnakul/SS; Peter Trusler/AG. **p.75**: Jenny Thynne/Flickr; Cre8design/SS. **p.76:** Sari ONeal/SS; Cathy Keifer/SS; Jenny Thynne/Flickr; Mongkolchon Akesin/SS; Justin Gilligan/AG. **p.77:** On the Wing Photography/SS; luchschenF/SS; Rose Colbeck/AG. **p.78**: Carloyn Smith 1/SS; Josephine Julian/SS; Rod Scott/AG; Laurent Renault/SS. **p.79:** Jenuine/AS; Eric Isselee/SS; Simia Attentive/SS. **p.80:** Christine Jones/@Mycohuman; Aastels/SS; Wanida_Sri/SS. **SECTION N: p.81:** Hendrickson Photography/SS. **p.82:** Wirestock/D; dwi putra stock/SS; Esther Beaton/AG. **p.83:** interestedbystandr/Flickr; Chris Davies/Wikimedia Commons; Andrew Gregory/AG. **SECTION O: p.84** Andrea Izzotti/SS. **p.85:** Wright Out There/SS; Justin Gilligan/AG; Brett Hondu/PB. **p.86:** Always Joy Photography/SS; Ken Griffiths/D;

Tom Grundy/SS. **SECTION P: p.87:** David Clode/US. **p.88** arturo nahum/SS; Andrei Stepanov/SS; samray/SS. **p.89** Kendall Collett/ SS; Rose Colbeck/AG;Jixiang Liu/D; Marje Crosby-Fairall/AG. **p.90** Jansen Chua/SS; Juergen_Wallstabe/ SS; Rose Colbeck/AG. **p.91** Andy Hutchinson/US; NASA/US. **p.92** (both) Anne Hayes/AG. **p.93:** Robert/ AS; New Africa/SS; Ego Guiotto/ AG. **p.94** Wang LiQiang/SS; Henri Koskinen/SS; Reddogs/AS; © Kevin Stead/AG. **p.95:** Andrey Burmakin/ SS; Phassa K/SS; Holger Hennern/ SS. **SECTION Q: p.96** Wirestock/D. **p.97:** T/Wikimedia Commons; NASA images/SS; AleksaStanko003/SS; Martin Pelanek/SS. **SECTION R**: **p.98** alybaba/SS. **p.99** mycteria/SS; Alex Cooper Photography/SS;Frances Lawlor/SS; oksankash/SS. **p.100** Peter Krisch/SS; Jean C Hebert/SS; IrinaK/ SS. **SECTION S: p.101** Wirestock/D. **p.102:** Esther Beaton/AG; Karen H Black/SS; Ekky Ilham/SS. **p.103:** crbellette/SS; Mamo studios/SS; KMT Professional Services/SS. **p.104:** kdshutterman/SS; IvaFoto/SS; Rose Marinelli/SS. **p.105:** Kichigin/SS. **SECTION T: p.106** Oleh Liubimtsev/ SS; Ego Guiotto/AG. **p.107:** (both) Ego Guiotto/AG. **p.108:** © Ego Guiotto/AG; Marje Crosby-Fairall/AG. **SECTION U: p.109** Alizada Studios/ SS; Ye.Maltsev/SS. **SECTION V: p.110** Graham Drew Photography/SS; Martin Pelanek/SS; Rose Colbeck/SS. **p.111:** David Clode/US; Ken Griffiths/ SS; fotoslaz/SS. **p.112:** Wirestock Creators/SS; Mike Dillon/AG; Andrew Sutton/SS. **p.113:** Ian Crocker/SS. **SECTION W: p.114:** Reality Images/ SS; Ondrej Prosicky/SS; Rose Colbeck/AG. **p. 115:** Don Fuchs/ AG; Willyam Bradberry/SS; Edward Stokes/AG. **p.116:** KPixMining/SS; Mitch Reardon/AG; Quentin Chester/ AG. **SECTION X: p.117**: photka/SS. **p.118:** ustas7777777/SS; Edgloris Marys/SS; SuradechKKPB/SS. **p.119** crbellette/SS; Gary D Chapman/SS. **p.120:** Harstela/ SS; Trish Jose/SS; Kuttelvaserova Stuchelova/SS. **SECTION Y: p.121:** Dslight photography/ SS; Arif Wicaksono/SS. **p.122:** EkaterinaKuzovkova/SS; BabyQ/ SS; Nicole Patience/SS. **SECTION Z: p.123**: Richard Lydekker/Wikimedia. **p.124** Jason Benz Bennee/SS Joshi Merbin/SS; Kevin Thiele/ Wikicommons. **p.125** Videologia/ SS; Marje Crosby Fairall/AG. **p.127:** Woraphon Banchobdi/D **p.128:** Ego Guiotto/AG; Anne Hayes/AG.

The GENUS name of the common seahorse (*Hippocampus kuda*) means 'bent horse'.

The red-rumped parrot (above) is ANISODACTYL.

Lemon ironbark (left) is an ANGIOSPERM.

First published in 2023

52–54 Turner Street
Redfern NSW 2016

editorial@ausgeo.com.au
australiangeographic.com.au

Author: Meghan Lindsay
Commissioning Editor: Karin Cox
Sub-Editors: Maggie Cooper, Serene Conneeley
Proofreader: Michele Perry
Creative Director: Aleksandra Beare
Designers: Chris Davies, Paul Hodge
Managing Picture Editor: Nicky Catley
Print Production: Andy Franks

AUSTRALIAN GEOGRAPHIC
Managing Director: David Haslingden
Licensing and Publishing Manager: Tom Bates
Commercial Assistant: Felicity McManus

Printed in China by 1010 Printing

A catalogue record for this book is available from the National Library of Australia

AUSTRALIAN GEOGRAPHIC SOCIETY
Enquiries about sponsorship and donations: 02 9136 7206
Email: society@ausgeo.com.au
www.australiangeographic.com.au/society

AUSTRALIAN GEOGRAPHIC SUBSCRIPTIONS
Sales and customer enquiries: 1300 555 176
australiangeographic.com.au